ANNALS OF MATHEMATICS STUDIES

Number 22

ANNALS OF MATHEMATICS STUDIES

Edited by Emil Artin and Marston Morse

1. Algebraic Theory of Numbers, *by* HERMANN WEYL
3. Consistency of the Continuum Hypothesis, *by* KURT GÖDEL
6. The Calculi of Lambda-Conversion, *by* ALONZO CHURCH
7. Finite Dimensional Vector Spaces, *by* PAUL R. HALMOS
10. Topics in Topology, *by* SOLOMON LEFSCHETZ
11. Introduction to Nonlinear Mechanics, *by* N. KRYLOFF *and* N. BOGOLIUBOFF
14. Lectures on Differential Equations, *by* SOLOMON LEFSCHETZ
15. Topological Methods in the Theory of Functions of a Complex Variable, *by* MARSTON MORSE
16. Transcendental Numbers, *by* CARL LUDWIG SIEGEL
17. Problème Général de la Stabilité du Mouvement, *by* M. A. LIAPOUNOFF
18. A Unified Theory of Special Functions, *by* C. A. TRUESDELL
19. Fourier Transforms, *by* S. BOCHNER *and* K. CHANDRASEKHARAN
20. Contributions to the Theory of Nonlinear Oscillations, *edited by* S. LEFSCHETZ
21. Functional Operators, Vol. I, *by* JOHN VON NEUMANN
22. Functional Operators, Vol. II, *by* JOHN VON NEUMANN
23. Existence Theorems in Partial Differential Equations, *by* DOROTHY L. BERNSTEIN
24. Contributions to the Theory of Games, *edited by* H. W. KUHN *and* A. W. TUCKER
25. Contributions to Fourier Analysis, *by* A. ZYGMUND, W. TRANSUE, M. MORSE, A. P. CALDERON, *and* S. BOCHNER
26. A Theory of Cross-Spaces, *by* ROBERT SCHATTEN
27. Isoperimetric Inequalities in Mathematical Physics, *by* G. POLYA *and* G. SZEGO

FUNCTIONAL OPERATORS

Volume II: The Geometry of Orthogonal Spaces

BY JOHN VON NEUMANN

PRINCETON
PRINCETON UNIVERSITY PRESS
1950

LONDON: GEOFFREY CUMBERLEGE, OXFORD UNIVERSITY PRESS

SECOND PRINTING, 1951

PRINTED IN THE UNITED STATES OF AMERICA

TABLE OF CONTENTS

FUNCTIONAL OPERATORS

CHAPTER XII.

LINEAR SPACES.

A space will now be considered whose elements f,g, ... are sometimes called points or vectors. It is assumed that an operation + is defined for every pair of elements of S and that multiplication on the left by a complex number is defined for each element of S.

Definition 12.1. *If f and g are any two elements of a space S, then S is called linear if f + g and af (a a complex number) are in S, and if*

a) $f + g = g + f$,

b) $(f + g) + h = f + (g + h)$,

c) $g + x = f$ *has at least one solution* x *in* S. (The uniqueness of this solution is not postulated, but it will be proved in Theorem 12.1.)

d) $a(bf) = (ab)f$,

e) $a(f + g) = af + ag$,

f) $(a + b)f = af + bf$,

g) $1f = f$.

Postulate A: *The space S is linear*

It will be assumed throughout that S satisfies Postulate A.

THEOREM 12.1. *If f and g are any two elements of S, then the equation $g + x = f$ has exactly one solution in S which will be denoted by $f - g$. $f - f$ is independent of f and will be denoted by 0.*

Proof: By Definition 12.1c, the equation $f + \zeta = f$ has a solution ζ and $f + x = g$ has a solution x. From the first of these equations it follows that $(f + x) + \zeta = (f + \zeta) + x = f + x$, so that, by the second equation, $g + \zeta = g$. Hence ζ is a solution of every equation $f + \zeta = f$. If η is another such solution, then $\eta + \zeta = \eta$, $\zeta + \eta = \zeta$, and by Definition 12.1a, $\zeta = \eta$. Hence every equation $f + \zeta = f$ has the unique solution ζ which will be denoted by 0, since confusion with the number zero is practically never a risk. Let z be a solution of $g + z = 0$. If $g + x = g + y$, then $(g + x) + z = (g + y) + z$, $(g + z) + x = (g + z) + y$, $0 + x = 0 + y$, and $x = y$. Hence $g + x = f$ has a unique solution x which will be denoted by $f - g$. It follows immediately that, for every element f, $f - f = 0$.

Definition 12.2. $0 - f$ will be denoted by $-f$.

It is obvious that $(f) + (-g) = (f) - (g)$.

THEOREM 12.2. For any element f of S, $(0)f = 0$, and for any complex number a, $a0 = 0$.

Proof: By Definition 12.1f, $(1)f = (1 + 0)f = (1)f + (0)f$. Since the equation $(1)f = (1)f + x$ has the unique solution $x = 0$, it follows that $(0)f = 0$. Again, by Definition 12.1e, $af = a(f + 0) = af + a0$, and $a0 = 0$.

THEOREM 12.3. For any element f of S and any complex number a, $(-a)f = a(-f) = -(af)$.

Proof: It follows from the preceding theorem that $0 = af + [-(af)]$, $0 = 0f = [a + (-a)]f = af + (-a)f$, and $0 = a0 = a[f + (-f)] = af + a(-f)$. The theorem follows from the fact that $0 = af + x$ has a unique solution.

Definition 12.3. If a complex number (f, g) is associated with each pair of elements f and g of S such that

a) $\overline{(f,g)} = (g,f)$ (*where* $\bar{z}$ *is the complex conjugate of* z),

b) $(f,f) > 0$ if $f \neq 0$,

c) $(af, g) = a(f,g)$ (*where* a *is a complex number*),

d) $(f_1 + f_2, g) = (f_1, g) + (f_2, g)$,

then (f,g) *is called the inner product of* f *and* g.

It follows from part a of this definition that (f, f) is real, so that the sign of inequality in part b has sense. If $f = g = 0$, then by part c, $(0, 0) = a(0, 0)$ for any complex number a. Hence $(0, 0) = 0$. Conversely, if $(f, f) = 0$, then f must be 0 to be consistent with part b. Hence $(f,f) = 0$ when and only when $f = 0$.

THEOREM 12.4. *If* (f, g) *is the inner product of* f *and* g, *then* $(g, af) = \bar{a}(g, f)$ *and* $(g, f_1 + f_2) = (g, f_1) + (g, f_2)$.

Proof: By Definition 12.3, $(g, af) = \overline{(af, g)} = \overline{a(f, g)} = \bar{a}\overline{(f, g)} = \bar{a}(g, f)$. Again, $(g, f_1 + f_2) = \overline{(f_1 + f_2, g)} = \overline{(f_1, g) + (f_2, g)} = \overline{(f_1, g)} + \overline{(f_2, g)} = (g, f_1) + (g, f_2)$.

Postulate B: An inner product is defined over the linear space S.

It will be assumed throughout that S satisfies Postulate B.

Definition 12.4. The positive square root $\sqrt{(f, f)}$ *is called the length of* f *and is denoted by* $\|f\|$.

It is obvious that $\|f\| = 0$ when and only when $f = 0$ and that $\|af\| = |a| \cdot \|f\|$.

THEOREM 12.5. (Schwarz's Lemma.) *If* f *and* g *are any two elements of* S, *then* $|(f, g)| \leq \|f\| \cdot \|g\|$.

Proof: Since $0 \leq (f-g, f-g) = (f,f) + (g,g) - (f,g) - (g,f) = \|f\|^2 + \|g\|^2 - 2\mathcal{R}(f,g)$, it follows that $\mathcal{R}(f,g) \leq \frac{1}{2}\|f\|^2 + \frac{1}{2}\|g\|^2$. If f and g are replaced by af and $\frac{1}{a}g$ (where $a > 0$), the left side of this inequality is unchanged, while the right side becomes $\frac{a^2}{2}\|f\|^2 + \frac{1}{2a^2}\|g\|^2$

The g.l.b. of this expression being $\|f\| \cdot \|g\|$, it follows that $\mathcal{R}(f, g) \leqq \|f\| \cdot \|g\|$. If f is replaced by θf (where $|\theta| = 1$), the left side of this last inequality becomes $\mathcal{R}\theta(f, g)$, while the right side remains unchanged. Since the maximum of $\mathcal{R}\theta(f, g)$ is $|(f, g)|$, therefore $|(f, g)| \leqq \|f\| \cdot \|g\|$.

If in the relation $(f - g, f - g) \geqq 0$ the sign of equality is to hold, it must be the case that $f = g$. If the relation $\mathcal{R}(f, g) \leqq \|f\| \cdot \|g\|$ is to be an equality, then $f = 0$ or $g = 0$ or $af = \frac{1}{a} g$, where a has the positive value which makes $\frac{a^2}{2} \|f\|^2 + \frac{1}{2a^2} \|g\|^2$ a minimum, that is, $f = \alpha g$, $\alpha > 0$. If the equality is to hold in the theorem itself, it must be the case that $f = 0$ or $g = 0$ or $\theta f = \alpha g$, where θ has the value which maximizes $\mathcal{R}\theta(f, g)$, that is, $f = \beta g$, β complex and $\neq 0$. Hence a necessary condition that the equality hold in the preceding theorem is that $f = 0$ or $g = 0$ or $f = \beta g$, $\beta \neq 0$. It is obvious that this condition is also sufficient.

<u>THEOREM 12.6.</u> <u>If</u> f <u>and</u> g <u>are any two elements of</u> S, <u>then</u> $\|f + g\| \leqq \|f\| + \|g\|$.

Proof: By Theorem 12.5, $(f + g, f + g) = \|f\|^2 + \|g\|^2 + 2\mathcal{R}(f, g) \leqq \|f\|^2 + \|g\|^2 + 2\|f\| \cdot \|g\|$. The theorem follows upon taking square roots.

In this theorem the equality holds if and only if $\mathcal{R}(f, g) = \|f\| \cdot \|g\|$, that is, as the discussion following Theorem 12.5 showed, if and only if $f = 0$ or $g = 0$ or $f = \alpha g$, $\alpha > 0$.

<u>Definition 12.5:</u> <u>If</u> f <u>and</u> g <u>are any two elements of</u> S, <u>then the distance between them is taken to be</u> $\|f - g\|$ <u>and is denoted by</u> D(f, g).

<u>THEOREM 12.7.</u> $D(f, f) = 0$, $D(f, g) > 0$ <u>when</u> $f \neq g$, $D(f, g) = D(g, f)$, $D(f, g) + D(g, h) \geqq D(f, h)$, $D(f + h, g + h) = D(f, g)$, <u>and</u> $D(af, ag) = |a| D(f, g)$.

The proofs of all parts of the theorem are apparent.

By Theorem 12.7 the distance $D(f, g)$ possesses all the properties which may be reasonably expected of a distance in a linear space. (For example, see F. Hausdorff, "Mengenlehre", W. de Gruyter and Co., Berlin and Leipzig, 1927, definitions on pp.94-97. For all topological questions, see the discussions, ibid., pp.94-138.) Hence $D(f, g)$ can be used to define a topology in S:

<u>Definition 12.6: By</u> $\lim_{n\to\infty} f_n = f$ <u>it is meant that</u> $\lim_{n\to\infty} \|f_n - f\| = 0$. <u>A function</u> $F(f)$ <u>is continuous at</u> f <u>if the condition</u> $\lim_{n\to\infty} f_n = f$ <u>implies the condition</u> $\lim_{n\to\infty} F(f_n) = F(f)$. (The definition domain of $F(f)$ must be contained in S, but its range of values may be contained in S or it may consist of complex numbers. This definition is of course equivalent to the statement that for every $\varepsilon > 0$ there exists a $\delta = \delta(f,\varepsilon) > 0$ such that whenever $\|g - f\| < \delta$ then $\|F(g) - F(f)\| \leqq \varepsilon$ or $|F(g) - F(f)| \leqq \varepsilon$.) The generalization of the notion of continuity to two or more variables (of which some may run over complex numbers) is apparent.

Let C be a subset of S. f is a condensation point of C if there exists a sequence $f_1, f_2, \ldots$ of elements of C such that $\lim_{n\to\infty} f_n = f$. (This is of course equivalent to the possibility of finding for each ε an element $g \in C$ such that $\|g - f\| < \varepsilon$.)

C is closed if it contains all its condensation points. C is dense in D if every point of D is a point of C or a condensation point of C. C is open if, for each $f \in C$, there exists an $\varepsilon = \varepsilon(f) > 0$ such that the entire sphere $\|g - f\| < \varepsilon$ is contained in C. (This means of course that the complement of C is closed. See Definition 1.8 et seq.)

Thus a topology is determined in S by means of the distance $D(f, g)$ and all topological problems relating to S can be discussed in the usual terminology.

THEOREM 12.8: The functions $f + g$, af, $-f$, $f - g$, (f, g), $\|f\|$ and $\|f - g\|$ are continuous in f, g, and a.

Proof: The relations

$$\|(f + g) - (f_o + g_o)\| = \|(f - f_o) + (g - g_o)\| \leq \|f - f_o\| + \|g - g_o\| ,$$

$$\|af - a_o f_o\| = \|(a_o + (a - a_o))(f_o + (f - f_o)) - a_o f_o)\| \leq$$

$$\leq |a_o| \cdot \|f - f_o\| + \|f_o\| \cdot |a - a_o| + |a - a_o| \cdot \|f - f_o\| ,$$

$$|(f, g) - (f_o, g_o)| = |(f_o + (f - f_o), g_o + (g - g_o)) - (f_o, g_o)| \leq$$

$$\leq \|f_o\| \cdot \|g - g_o\| + \|g_o\| \cdot \|f - f_o\| + \|f - f_o\| \cdot \|g - g_o\|$$

show that $f + g$, af, and (f, g) are continuous. If $a = -1$, then af becomes $-f$; if g is replaced by $-g$, then $f + g$ becomes $f - g$; if $g = f$, then $\sqrt{(f,g)} = \|f\|$; if f is replaced by $f - g$, then $\|f\|$ becomes $\|f - g\|$.

The theory of spaces S has been based so far on the notions of $f + g$ and af (linearity) and (f, g) (inner product). $\|f\|$ and "distance" were defined in terms of these notions and their essential properties determined by means of them. But topological literature is much more familiar with the following set of primitive notions: $f + g$ and af (linearity), and $\|f\|$ (metricity). (cf. Hausdorff, loc. cit.) Then the fundamental properties of $\|f\|$ (mentioned after Definition 12.4 and in Theorem 12.6) are to be postulated, while (f, g) will be defined in terms of the metric and its properties proved. But to do this, additional postulates concerning $\|f\|$ are needed. This situation will now be discussed in detail.

Definition 12.7: A linear space S is called metric if an absolute value $\|f\|$ is defined in S such that

a) $\|f\| > 0$ if $f \neq 0$,

b) $||af|| = |a| \cdot ||f||$,

c) $||f + g|| \leq ||f|| + ||g||$.

(Then of course $D(f, g) = ||f - g||$ defines a distance for which Theorem 12.7 is valid.)

Postulate B_1: The linear space S is metric.

It will be assumed for the moment that S satisfies Postulates A and B_1, but not necessarily B.

THEOREM 12.9: If $C(f, g) = \frac{1}{2} \frac{||f + g||^2 + ||f - g||^2}{||f||^2 + ||g||^2}$ (where f and g are in S and not both 0), and if α = g.l.b. of $C(f, g)$ and β = l.u.b. of $C(f, g)$ for all f and g in S, then $\frac{1}{2} \leq \alpha \leq 1 \leq \beta \leq 2$ and $\alpha\beta = 1$.

Proof: First, $C(f + g, f - g) = \frac{1}{2} \frac{4||f||^2 + 4||g||^2}{||f + g||^2 + ||f - g||^2} = \frac{1}{C(f, g)}$.

Hence if $C(f, g)$ ever assumes a value a, it also assumes the value $\frac{1}{a}$. Therefore $\alpha\beta = 1$ and, since $\alpha \leq \beta$, $\alpha \leq 1 \leq \beta$. Second, $||f + g||^2 + ||f - g||^2 \leq$ $\leq 2(||f|| + ||g||)^2 \leq 4(||f||^2 + ||g||^2)$, so that $C(f, g) \leq 2$. Thus $\beta \leq 2$, and since $\alpha\beta = 1$, $\alpha \geq \frac{1}{2}$. This completes the proof.

The following examples show that α, β may assume all values compatible with Theorem 12.9; that is, that β may assume all values ≥ 1, ≤ 2. Let S be the real Euclidean space with rectangular coordinates. If f is the point with coordinates x_1, x_2, let $||f||_p = (|x_1|^p + |x_2|^p)^{\frac{1}{p}}$, where p is some fixed number ≥ 1. In case $p = \infty$, we may interpret $||f||_p = \text{Max}(|x_1|, |x_2|)$. $\beta = \beta_p$ is obviously a continuous function of p. For $p = 1$ and $p = \infty$ the choices $f = (1, 0)$, $g = (0, 1)$ resp. $f = (1, 1)$, $g = (1, -1)$ prove that $\beta_p = 2$. For $p = 2$, the following Theorem 12.10 (or an easy direct verification) will give $\beta_p = 1$. Thus all values ≥ 1, ≤ 2 are assumed by the β.

Postulate B_2: $C(f, g) = 1$.

THEOREM 12.10: In a linear metric space S, Postulate B_2 is the neces-

sary and sufficient condition that an inner product (f, g) satisfying Postulate B may be defined in such a way that $\|f\| = \sqrt{(f, f)}$. If (f, g) can be so defined, this can be done in only one way.

All this remains true if the condition b) in Definition 12.7 is replaced by the weaker condition

b') $$\lim_{a \to 0} \|af\| = 0 \quad \textit{and} \quad \|if\| = \|f\| .$$

Proof: If there exists an inner product (f, g) such that $\|f\| = \sqrt{(f, f)}$, then

$$(f + g, f + g) + (f - g, f - g) = 2(f, f) + 2(g, g),$$

so that $\|f + g\|^2 + \|f - g\|^2 = 2\|f\|^2 + 2\|g\|^2$ and C(f, g) = 1. Thus Postulate B_2 is necessary. Again

$$(f + g, f + g) - (f - g, f - g) = 2(f, g) + 2(g, f) = 4\mathcal{R}(f, g).$$

Considering $\mathcal{J}(f, g) = -\mathcal{R}i(f, g) = -\mathcal{R}(if, g)$, we have

$$\mathcal{R}(f, g) = \tfrac{1}{4}(\|f + g\|^2 - \|f - g\|^2),$$
(*)
$$(f, g) = \mathcal{R}(f, g) - i\mathcal{R}(if, g).$$

Hence (f, g) if it exists and is such that $\|f\| = \sqrt{(f, f)}$, is uniquely determined by $\|f\|$. (Observe that so far Postulate B_1, that is Definition 12.7, has not been used at all.)

It remains to prove the sufficiency of Postulate B_2. Suppose the space S satisfies Postulates **A**, B_1 and B_2, but with b') instead of b) in Definition 12.7. Define (f, g) by the equations (*). It must be shown that (f, g) satisfies the conditions in Definition 12.3, and that $\|f\| = \sqrt{(f, f)}$.

Condition b') implies $\|0\| = 0$, and so C(f, g) = 1 gives for f = 0, $\|g\| = \|-g\|$. This implies $\mathcal{R}(0, g) = 0$.

Replace now in the first part of (*) f by $f_1 \pm f_2$, and add. Using C(f, g) = 1, that is $\|k + h\|^2 + \|k - h\|^2 = 2\|k\|^2 + 2\|h\|^2$, this gives

$\mathcal{R}(f_1+ f_2, g) + \mathcal{R}(f_1- f_2, g) =$

$= \frac{1}{4}\{\|f_1+ f_2+ g\|^2-\|f_1+ f_2- g\|^2+\|f_1- f_2+ g\|^2-\|f_1- f_2- g\|^2\} =$

$= \frac{1}{4}\{(\|f_1+ g + f_2\|^2+\|f_1+ g - f_2\|^2) - (\|f_1- g + f_2\|^2+\|f_1- g - f_2\|^2)\} =$

$= \frac{1}{4}\{(2\|f_1+ g\|^2+ 2\|f_2\|^2) - (2\|f_1- g\|^2+ 2\|f_2\|^2)\} =$

$= \frac{1}{2}\{\|f_1+ g\|^2- \|f_1- g\|^2\} = 2\mathcal{R}(f_1, g).$

Put $f_1 = f_2$, then there results $\mathcal{R}(2f_1, g) = 2\mathcal{R}(f_1, g)$. Hence we have $\mathcal{R}(f_1+ f_2, g) + \mathcal{R}(f_1- f_2, g) = \mathcal{R}(2f_1, g)$, or, replacing f_1, f_2 by $\frac{f_1+ f_2}{2}$, $\frac{f_1- f_2}{2}$; $\mathcal{R}(f_1, g) + \mathcal{R}(f_2, g) = \mathcal{R}(f_1+ f_2, g)$. Now replace f_1, f_2 by if_1, if_2, and use the second part of (*):

(‡) $$(f_1, g) + (f_2, g) = (f_1+ f_2, g).$$

Let us now discuss equation

(‡‡) $$(af, g) = a(f, g).$$

Denote by S the set of all (complex) a for which it holds. (‡) implies that a, $b \in S$ imply $a \pm b \in S$. Obviously $1 \in S$; so all integers $0, \pm 1, \pm 2, \ldots$ belong to S. As $a, b \in S$, $b \neq 0$ obviously imply $\frac{a}{b} \in S$, all rational numbers belong to S.

By Definition 12.7, c), $\|h\| - \|k\| \leqq \|h - k\|$ (replace f, g by h - k, k). Interchanging h, k, we get $\|k\| - \|h\| \leqq \|h - k\|$; thus $\big|\|h\| - \|k\|\big| \leqq \|h - k\|$. Hence $\big|\|\alpha f + g\| - \|\beta f + g\|\big| \leqq \|(\alpha - \beta)f\|$, and if $\alpha \to \beta$, this converges to 0 by b'). So $\|\alpha f + g\|$ is a continuous function of α ; similarly $\|\alpha f - g\|$. Therefore $\mathcal{R}(\alpha f, g)$ and $(\alpha f, g)$ are by (*) also continuous functions of α . Thus S is a closed set. As the real α's are limits of rational α's, all real α's belong to S.

Now (‡‡) holds for $\alpha = i$, as follows directly from (*). Hence $i \in S$, and if α_1, α_2 are real, $\alpha_1 - i\alpha_2 = \alpha_1 + \frac{\alpha_2}{i} \in S$. Thus all complex α's belong to S; that is (‡‡) always holds.

We have verified the conditions c), d) of Definition 12.3, a) mean $\mathcal{R}(f, g) = \mathcal{R}(g, f)$, $\mathcal{R}(if, g) = -\mathcal{R}(ig, f)$. The first relation follows directly from (*). As for the second, observe that $\|if\| = \|f\|$ and (*) give $\mathcal{R}(if, ig) = \mathcal{R}(f, g)$; hence $\mathcal{R}(if, g) = \mathcal{R}(-f, ig) = -\mathcal{R}(f, ig) = -\mathcal{R}(ig, f)$.

It remains to prove b) and $\|f\| = \sqrt{(f, f)}$, but the former follows from the latter. This means $\mathcal{R}(f, f) = \|f\|^2$, $\mathcal{R}(if, f) = 0$. Now a), c) imply $(f, \alpha g) = \bar{\alpha}(f, g)$, and so $(\alpha f, \alpha f) = |\alpha|^2(f, f)$. Therefore, $(2f, 2f) = 4(f, f)$, $((1 + i)f, (1 + i)f) = ((-1 + i)f, (-1 + i)f) = 2(f, f)$, proving our statements.

Thus the spaces satisfying Postulates A and B are identical with those satisfying A, B_1, and B_2, that is, with those linear metric spaces in which the invariants α and β defined in Theorem 12.9 assume their extreme values $\alpha = \beta = 1$. Each of the notions (f, g) and $\|f\|$ can be used to derive the other.

It is now desirable to resume the investigation of the properties of spaces S satisfying Postulates A and B.

Definition 12.8: If f and g are any two elements of S, they are called orthogonal, $f \perp g$, if $(f, g) = 0$; f is called normalized if $\|f\| = 1$. A set A of elements f, g, ... in S is called orthogonal if each pair of distinct elements of A are orthogonal; A is called normalized if each element of A is normalized. If A is orthogonal and normalized, it is called ortho-normal (o.n.); if A is o.n., it is called complete or maximal in S if it is not a proper part of any other o.n. set in S.

It is obvious that if A is o.n., then each subset of A is also o.n.

If f and g are any two elements of an o.n. set A, then $(f, g) = \begin{cases} 1 \text{ if } f = g, \\ 0 \text{ if } f \neq g. \end{cases}$ Conversely, if A is such a set that any two of its elements satisfy this condition, then A is o.n. Thus if A consists of a finite or infinite sequence $\varphi_1, \varphi_2, \ldots$ of distinct elements, the o.n. character of A is equivalent to the condition $(\varphi_m, \varphi_n) = \delta_{mn}$, where δ_{mn} is the well-known Kronecker symbol $\delta_{mn} = \begin{cases} 1 \text{ if } m = n, \\ 0 \text{ if } m \neq n. \end{cases}$ In the sequel, the letters $\varphi, \psi, \ldots$ will always be used to denote the elements of an o.n. set.

That an o.n. set A is complete in S means that there exists no element $\varphi \in S$ such that the set consisting of A and φ is o.n.; that is, there exists no normalized element $\varphi \in S$ such that $\varphi \perp A$. But if $f \neq 0$ and $f \perp A$, where $f \in S$, then $\varphi = \frac{1}{\|f\|} f$ is normalized and orthogonal to A. Thus it follows that the completeness in S of an o.n. set A in S is equivalent to the fact that the only element in S orthogonal to A is $f = 0$.

If $\varphi_1, \varphi_2, \ldots$ is an o.n. set of elements and if $g = \sum_{i=1}^{n} x_i \varphi_i$, then it is obvious that $x_j = (g, \varphi_j)$. If $\sum_{i=1}^{\infty} f_i$ is defined to be $\lim_{n\to\infty} \sum_{i=1}^{n} f_i$ (provided that this limit exists), and if $g = \sum_{i=1}^{\infty} x_i \varphi_i$, then $x_j = (g, \varphi_j)$ since (p, q) is continuous and since $x_j = \lim_{n\to\infty} (\sum_{i=1}^{n} x_i \varphi_i, \varphi_j) = (\sum_{i=1}^{\infty} x_i \varphi_i, \varphi_j)$.

<u>THEOREM 12.11: If $\varphi_1, \ldots, \varphi_n$ is a finite o.n. set in S and if $g \in S$, then $V = \|g - \sum_{i=1}^{n} x_i \varphi_i\|$ is minimized when $x_j = (g, \varphi_j)$; if V_{min} is the minimum value, then $0 \leq V_{min}^2 = \|g\|^2 - \sum_{i=1}^{n} |(g, \varphi_i)|^2$, so that $\|g\|^2 \geq$</u> $\geq \sum_{i=1}^{n} |(g, \varphi_i)|^2$. (This last relation is known as Bessel's inequality.)

Proof: $0 \leq \|g - \sum_{i=1}^{n} x_i \varphi_i\|^2 = (g - \sum_{i=1}^{n} x_i \varphi_i, g - \sum_{i=1}^{n} x_i \varphi_i) =$

$= (g,g) - (g, \sum_{i=1}^{n} x_i \varphi_i) - (\sum_{i=1}^{n} x_i \varphi_i, g) + (\sum_{i=1}^{n} x_i \varphi_i, \sum_{j=1}^{n} x_i \varphi_i) =$

$$= (g, g) - \sum_{i=1}^{n} \bar{x}_i(g, \varphi_i) - \sum_{i=1}^{n} x_i(\varphi_i, g) + \sum_{i,j=1}^{n} (x_i \varphi_i, x_j \varphi_j) =$$

$$= (g, g) - 2R[\sum_{i=1}^{n} x_i(\overline{g, \varphi_i})] + \sum_{i=1}^{n} |x_i|^2 =$$

$$= \left\{\sum_{i=1}^{n} [|x_i|^2 - 2R\{x_i(\overline{g, \varphi_i}) + |(g, \varphi_i)|^2]\right\} + ||g||^2 - \sum_{i=1}^{n} |(g, \varphi_i)|^2 =$$

$$= (\sum_{i=1}^{n} |x_i - (g, \varphi_i)|^2) + ||g||^2 - \sum_{i=1}^{n} |(g, \varphi_i)|^2.$$

The theorem follows from the form of this last expression.

Corollary 1: If A is any o.n. set in S, then Bessel's inequality holds for any finite subset of A. If A is an infinite sequence $\varphi_1, \varphi_2, \ldots$, then the series $\sum_{i=1}^{\infty} |(g, \varphi_i)|^2$ is convergent and $\leqq ||g||^2$.

The statements of the following Corollaries 2, 3, and 4 are of interest only if A is non-countable.

Corollary 2: If A is any o.n. set in S, then $(g, \varphi) = 0$ for every φ in A except for a countable subset and $\sum_{\varphi \in A} |(g, \varphi)|^2$ not only has sense but is convergent and $\leqq ||g||^2$. (This last relation, or that one in Corollary 1, is again Bessel's inequality; if it is actually an equality it is called Parseval's equation.)

Corollary 3: If A is any o.n. set in S, and if $g_1, g_2, \ldots$ is any sequence of elements of S, then all but a countable set of φ's in A are simultaneously orthogonal to all the g's.

Proofs: Corollary 1 is apparent. It follows from Bessel's inequality that the number k of φ's such that $|(g, \varphi)|^2 > \varepsilon$ is not greater than $\frac{||g||^2}{\varepsilon}$; thus k is finite. If ε is given the values 1, 1/2, 1/3, ..., the corresponding sets of φ's such that $|(g, \varphi)|^2 > \varepsilon$ make up a countable set and comprise all φ's such that $|(g, \varphi)|^2 > 0$. This proves Corollary 2. Corollary

3 follows at once.

Definition 12.9. <u>A set T is called separable if there exists a countable set of its elements which is dense in T.</u>

Corollary 4. <u>If A is any o.n. set in S, and if T is a separable part of S, then all the elements of A except for a countable subset are simultaneously orthogonal to all the elements of T.</u>

Proof: Let $f_1, f_2, \ldots$ be a sequence $\sum$ of elements of T dense in T. Let g be any element of T and let $\{f_{g_i}\}$ be a subsequence of $\sum$ such that $\lim_{n\to\infty} f_{g_n} = g$. Except for a countable set of φ's in A, $(\varphi, f_i) = 0$ for all i and for all φ in A. Hence, with at least the same exceptions, $(\varphi, f_{g_i}) = 0$ for all i and (by continuity) $(\varphi, g) = 0$.

Corollary 5: <u>If A is any o.n. set in S, and if S itself is separable, then A is countable.</u>

Proof: If A were not countable it would contain elements φ which would be orthogonal to the whole of S. Such an element φ would therefore be orthogonal to itself. Hence $\varphi = 0$. This contradicts the normalized character of A.

THEOREM 12.12: <u>If S is separable, any subspace S' of S is also separable.</u>

Proof: Suppose the sequence $\sum : f_1, f_2, \ldots$ is in S and dense in S. Corresponding to each pair of positive integers m and n let an element g_{mn} of S' be selected such that $D(g_{mn}, f_n) < \frac{1}{m}$ (providing, of course, that there exists such an element). Then the elements g_{mn} are dense in S'. To show this, let g be an element of S'. Then there exists a subsequence $f_{i_1}, f_{i_2}, \ldots$ of $\sum$ such that $\lim_{n\to\infty} D(g, f_{i_n}) = 0$. For sufficiently large n, $D(g, f_{i_n}) < \frac{1}{m}$, and then elements $h \in S'$ with $D(h, f_{i_n}) < \frac{1}{m}$ exist; for example, $h = g$. Thus g_{mi_n} is de-fixed $D(g, f_{i_n}) < \frac{1}{m}$, $D(g_{mi_n}, f_{i_n}) < \frac{1}{m}$, and therefore $D(g, g_{mi_n}) < \frac{2}{m}$ by Theorem 12.6.

Definition 12.10: *A finite set of elements $f_1, \dots, f_n$ of S is called linearly independent if the condition $\sum_{i=1}^{n} x_i f_i = 0$ is satisfied only by the set of values $x_1 = \dots = x_n = 0$.*

It is easily seen that every finite o.n. set $\varphi_1, \dots, \varphi_n$ is linearly independent, for the condition $\sum_{i=1}^{n} x_i \varphi_i = 0$ implies (by the remark before Theorem 12.11) that $x_j = (0, \varphi_j) = 0$.

THEOREM 12.13. *If $f_1, \dots, f_n$ are linearly independent elements of a space S, then there exists an o.n. set of elements $\varphi_1, \dots, \varphi_n$ in S.*

Proof: Since $f_1 \neq 0$ it can be written $f_1 = a_1 \varphi_1$, where $a_1 \neq 0$ and $\|\varphi_1\| = 1$. Since f_1 and f_2 are linearly independent, the difference $f_2 - (f_2, \varphi_1)\varphi_1$ is different from zero, is orthogonal to φ_1, and can be written as $a_2 \varphi_2$, where $a_2 \neq 0$, $\|\varphi_2\| = 1$. φ_2 is orthogonal to φ_1. Since f_1, f_2, and f_3 are linearly independent, the difference $f_3 - \sum_{i=1}^{2}(f_3, \varphi_i)\varphi_i$ is different from zero, is orthogonal to φ_1 and φ_2, and can be written as $a_3 \varphi_3$, where $a_3 \neq 0$, $\|\varphi_3\| = 1$. φ_3 is orthogonal to φ_1 and φ_2. This process may be continued until the set $\varphi_1, \dots, \varphi_n$ is obtained.

A sequence $f_1, f_2, \dots$ of elements of S is called fundamental if $\lim_{m,n\to\infty} D(f_m, f_n) = 0$ (that is, if $\lim_{m,n\to\infty} (f_m - f_n) = 0$).

Definition 12.11. *A space S is called complete if, corresponding to each fundamental sequence $f_1, f_2, \dots$ of elements of S there exists an element f of S such that $\lim_{n\to\infty} D(f, f_n) = 0$ (that is, such that $\lim_{n\to\infty} f_n = f$).*

The following alternatives holdfor a space S: 1) if all finite sets Σ_n of linearly independent elements of S are formed, where n is the number of elements in a particular set Σ_n, the least upper bound N of n may be finite or infinite. By Theorem 12.13 and the remark following Definition 12.10 this alternative may be replaced by the alternative that the least upper bound N of n with respect to all finite o.n. sets of elements of S is finite or infinite. 2) S may or may not be separable. 3) S may or may not be complete.

THEOREM 12.14: *If S is such that N (of the previous paragraph) is finite, then S can be represented by the complex N-dimensional Euclidean space; hence S is separable and complete.*

Proof: There exists an o.n. set $\varphi_1, \ldots, \varphi_N$ in S. It is apparent that any element f of S can be represented as $f = \sum_{i=1}^{N} x_i \varphi_i$ in one and only one way. If E_N is the complex N-dimensional Euclidean space, then f can be represented in E_N by the point $(x_1, \ldots, x_N)$. Similarly, any other element g of S would be represented by the point $(y_1, \ldots, y_N)$. It follows that $(f + g) \sim (x_1 + y_1, \ldots, x_N + y_N)$ and $af \sim (ax_1, \ldots, ax_N)$, so that f + g and af are in E_N. Conversely, since S is linear, there is an element of S corresponding to each point of E_N. Furthermore, $(f, g) = (\sum_{i=1}^{N} x_i \varphi_i, \sum_{j=1}^{N} y_j \varphi_j) = \sum_{i=1}^{N} x_i \bar{y}_i$, so that S is isomorphic with E_N.

If S is such that N is finite, then an o.n. set $\varphi_1, \ldots, \varphi_k$ is complete when and only when k = N.

Definition 12.12. *A space S is said to be extended to a space T if T is obtained from S by the addition of elements to S, and if f + g, af, and (f, g) are defined over T in such a manner that, for elements of S, the definitions of these notions in S and T agree.* (All these spaces are assumed to satisfy Postulates A and B.)

THEOREM 12.15. *Any space S can be extended to a complete space $\mathfrak{T}$ in essentially only one way such that S is dense in $\mathfrak{T}$. S and $\mathfrak{T}$ have the same value of N and the same character with regard to separability.*

Proof: Let $F = \{f_i\}$ be a fundamental sequence of elements of S. F is said to be equivalent to G, $F \sim G$, if $\lim_{n\to\infty} D(f_n, g_n) = 0$. It follows that $F \sim F$; if $F \sim G$, $G \sim F$; if $F \sim G$ and $G \sim H$, $F \sim H$. F will now be regarded as an element of a space $\mathcal{T}$, and all fundamental sequences equivalent to F will be regarded as the same element of $\mathcal{T}$. If $\{f_i\}$ and $\{g_i\}$ are fundamental sequences, then $\{f_i + g_i\}$ and $\{af_i\}$ are also and the operations $F + G$ and aF in $\mathcal{T}$ are defined to be $\{f_i + g_i\}$ and $\{af_i\}$; (F, G) is taken to be $\lim_{n\to\infty} (f_n, g_n)$. That this limit exists follows from the fact that $|(f_n, g_n) - (f_m, g_m)| = |(f_m + (f_n - f_m), g_m + (g_n - g_m)) - (f_m, g_m)| \leq \|f_n - f_m\| \cdot \|g_m\| + \|f_m\| \cdot \|g_n - g_m\| + \|f_n - f_m\| \cdot \|g_n - g_m\|$ and $\lim_{m,n\to\infty} [(f_n, g_n) - (f_m, g_m)] = 0$ since $\|f_m\|$ and $\|g_m\|$ are bounded (insomuch as $\|f_m\| = \|f_n + (f_m - f_n)\| \leq \|f_n\| + \|f_m - f_n\| \leq \|f_n\| + \varepsilon$ for a sufficiently great but fixed n and all $m > n$). It is apparent that sequences equivalent to $\{f_i\}$ and $\{g_i\}$ may be used in the definitions of $F + G$, aF, and (F, G) without doing more to the first two of these entities than replacing them by equivalent entities and without changing the last entity at all. Hence the same operations occur in S and $\mathcal{T}$. If $F_f = \{f, f, \ldots\}$ and $F_g = \{g, g, \ldots\}$, then $F_f = F_g$ when and only when $f = g$. Hence f, F_f and all fundamental sequences equivalent to F_f may be identified, and if there exists an element f corresponding to a given element F, operations on f and F have the same significance. Let $\mathcal{T}$ be the space of all fundamental sequences arising from S. Then $\mathcal{T}$ has the properties of S in that it is linear with an inner product, as all parts of Definitions 12.1 and 12.3 can be easily verified. (In particular, the condition $(F, F) = 0$ implies that $F \sim 0$, and thus in the present terminology $F = 0$.) Finally, S is isomorphic to, and in the present terminology, even identical with a part of $\mathcal{T}$: the set of all F_f.

By Theorem 12.14, if N is finite for S, S is complete and therefore

$S = \mathscr{T}$. If N is infinite for S, it is infinite for $\mathscr{T}$. Hence S and $\mathscr{T}$ have the same character with regard to the first of the alternatives mentioned before Theorem 12.14 and they have the same value of N.

If F belongs to $\mathscr{T}$, then F is a fundamental sequence $\{f_i\}$, $\lim_{i,j\to\infty} \| f_i - f_j \| = 0$, so that $\lim_{i\to\infty} (\lim_{j\to\infty} \| f_i - f_j \|) = 0$. Now $D(f_i, F) = $ $= \lim_{j\to\infty} \| f_i - f_j \|$. Therefore $\lim_{i\to\infty} D(f_i, F) = 0$, that is, $\lim_{i\to\infty} f_i = F$. Hence S is dense in $\mathscr{T}$.

If $\mathscr{T}$ is separable, then its subset S is also; if S is separable, $\mathscr{T}$ is also, since S is dense in $\mathscr{T}$. Hence S and $\mathscr{T}$ have the same character with regard to the second of the above mentioned alternatives.

$\mathscr{T}$ is always complete, for let F_1, F_2, ... be any fundamental sequence in $\mathscr{T}$. Since S is dense in $\mathscr{T}$, there exists an element $\mathbf{f_n}$ of S such that $D(F_n - f_n) < \frac{1}{n}$. The sequence $\{f_n\}$ is fundamental since $\| f_m - f_n \| = $ $= \| (f_m - F_m) + (F_m - F_n) + (F_n - f_n) \| \leq \frac{1}{m} + \frac{1}{n} + \| F_m - F_n \|$. Hence $\{f_n\}$ is an element F of $\mathscr{T}$ such that $\lim_{n\to\infty} D(f_n, F) = 0$, so that $\lim_{n\to\infty} D(F_n, F) = 0$. This completes the proof except for the uniqueness of $\mathscr{T}$.

Suppose there were a second extension $\mathscr{T}'$ of S such that S is dense in $\mathscr{T}'$. If $\{f_i\}$ is a fundamental sequence in S, and therefore an element of $\mathscr{T}$, then $\{f_i\}$ has a limit φ in $\mathscr{T}'$. Likewise the fundamental sequence $\{g_i\}$ has a limit ψ in $\mathscr{T}'$. But φ and ψ cannot be equal unless $\{f_i - g_i\}$ converges to 0, that is, unless $\{f_i\}$ and $\{g_i\}$ are equivalent. Hence distinct elements of $\mathscr{T}$ correspond to distinct elements of $\mathscr{T}'$, and conversely, distinct elements of $\mathscr{T}'$ correspond to distinct elements of $\mathscr{T}$. Since $\mathscr{T}'$ is complete, this mapping covers all of $\mathscr{T}$. Since S is dense in $\mathscr{T}'$, this mapping provides an image in $\mathscr{T}$ for every element of $\mathscr{T}'$. Hence the mapping is one-to-one between $\mathscr{T}$ and $\mathscr{T}'$. It is obvious that the operations F + G, aF, and (F, G) have the significance in $\mathfrak{H}'$ as in $\mathfrak{H}$, so that the mapping is isomorphic.

It is now possible to consider the following postulates concerning the space S:

<u>A</u>. S is linear. (Definition 12.1.)

<u>B</u>. An inner product is defined over S. (Definition 12.3.)

$\underline{C_1}$. N is finite.
$\underline{C_2}$. N is infinite.
} (Alternative 1 after Definition 12.11.)

$\underline{D_1}$. S is separable.
$\underline{D_2}$. S is not separable.
} (Definition 12.9, alternative 2 after Definition 12.11.

$\underline{E_1}$. S is complete.
$\underline{E_2}$. S is not complete.
} (Definition 12.11, alternative 3 after Definition 12.11

There are, a priori, eight possible combinations of the postulates A, B, C_ρ, D_σ, E_τ $(\rho, \sigma, \tau = 1, 2)$ for S. But by Theorem 12.14, C_1 implies D_1 and E_1, and that S is the complex N-dimensional Euclidean space. Thus only the four combinations of C_2 with D_σ and E_τ remain for consideration. By Theorem 12.15 it is possible to replace S by a complete space $\mathfrak{H}$ without altering any of the other conditions. Hence it may be assumed that E_1 holds.

Thus, besides the combination A, B, C_1 (and D_1, E_1), the only combinations to be considered are A, B, C_2, D_1 or D_2, and E_1. These latter cases will be the main concern of these investigations. In any event, it will be assumed throughout the sequel that S satisfies Postulates A, B, and E_1.

<u>THEOREM 12.16</u>: <u>If $\varphi_1, \varphi_2, \ldots$ is an o.n. set in S, then a necessary and sufficient condition that $\sum_{i=1}^{\infty} a_i \varphi_i$ be convergent is that $\sum_{i=1}^{\infty} |a_i|^2$ be convergent.</u>

Proof: The condition is necessary, for if $\sum_{i=1}^{\infty} a_i \varphi_i$ is convergent, then $\lim_{\substack{m,n\to\infty \\ m>n}} \left\|\sum_{i=1}^{m} a_i \varphi_i - \sum_{i=1}^{n} a_i \varphi_i\right\| = 0$, and therefore

$$= \lim_{\substack{m,n\to\infty \\ m>n}} \left\| \sum_{i=1}^{m} a_i\varphi_i - \sum_{i=1}^{n} a_i\varphi_i \right\|^2 = \lim_{\substack{m,n\to\infty \\ m>n}} \left\| \sum_{i=n+1}^{m} a_i\varphi_i \right\|^2 =$$

$$\lim_{\substack{m,n\to\infty \\ m>n}} \left(\sum_{i=n+1}^{m} a_i\varphi_i, \sum_{j=n+1}^{m} a_j\varphi_j \right) = \lim_{\substack{m,n\to\infty \\ m>n}} \sum_{i=n+1}^{m} |a_i|^2 = 0.$$ Hence $\sum_{i=1}^{\infty} |a_i|^2$ is convergent. The condition is sufficient by a reversal of the argument in view of the fact that S is complete and that $\lim_{\substack{m,n\to\infty \\ m>n}}$ can be replaced in this case by $\lim_{m,n\to\infty}$.

Corollary: If A is any o.n. set in S, then $\sum_{\varphi\in A} (f, \varphi)\varphi$ is convergent, where f is any element of S.

Proof: This follows immediately from Corollary 2 of Theorem 12.11 together with Theorem 12.16.

THEOREM 12.17: If A is any o.n. set in S, if f is any element in S, and if g is defined by the condition $f = \sum_{\varphi\in A} (f, \varphi)\varphi + g$, then $g \perp A$.

Proof: If A is countable, then all the elements of A may be included in the summation and, for any φ_j of A, $(f, \varphi_j) = (\sum_{i=1}^{\infty} (f, \varphi_i)\varphi_i, \varphi_j) + (g, \varphi_j) = (f, \varphi_j) + (g, \varphi_j)$ and $(g, \varphi_j) = 0$. If A is not countable, consider a given $\varphi^0 \in A$. It is sufficient to extend the summation $\sum_{\varphi\in A}$ over those $\varphi \in A$ for which $(f, \varphi) \neq 0$ and, in addition, φ^0 (which may or may not be such that $(f, \varphi^0) \neq 0$). All these φ form a sequence, and for them the condition $(g, \varphi) = 0$ obtains; in particular, $(g, \varphi^0) = 0$. This completes the proof since φ^0 was arbitrary.

Definition 12.13: Let M be a subset of S. If $f + g \in M$ whenever $f \in M$ and $g \in M$, and if $af \in M$ whenever $f \in M$, then M is called linear.

It is clear that M is complete in the sense of Definition 12.11 if and only if it is closed in the topological sense (Definition 12.6) inasmuch as S is complete.

It is obvious that the product of any set of linear (closed) sets is linear (closed). A linear set in S is usually called a linear manifold (l.m.) in S.

If X is any set in S, there exists a l.m. M in S containing X (for example, S itself). The common part of all l.m.'s containing X is the "smallest" l.m. containing X; it will be called "the" l.m. determined by X and will be denoted by $\{X\}$. The symbol [X] is defined similarly with respect to closed linear manifolds (c.l.m.). In general, $\{X, Y, \ldots, f, g, \ldots\}$ is the l.m. determined by the sum of X, Y, ..., f, g, ..., and [X, Y, ..., f, g, ...] is the c.l.m. defined analogously. It is obvious that $[X] \supset \{X\} \supset X$.

If X is any set in S, then $\{X\}$ and [X] can be constructed directly in the following manner, as is easily verified: $\{X\}$ is the set of all elements $\sum_{i=1}^{n} a_i f_i$, where $f_i \in X$ and $n = 1, 2, \ldots$. [X] is the set of all condensation points of $\{X\}$. $\{X\} = [X]$ if X has a finite basis.

It is now desirable to take up the key theorem of the present general discussion of spaces S: The existence of a complete o.n. set in any space S. This theorem holds without any further restrictions on S, but in order to prove it, it is necessary to use rather deep results of the general theory of sets; the so-called "well ordering theorem" of Zermelo, G. Cantor's "transfinite ordinal numbers", and the possibility of definition by "transfinite induction". (For a systematic exposition of this theory, see, for example, Hausdorff, loc. cit. before Definition 12.6, pp.55-58, 58-62, and 62 respectively. For the notion of "power" and "equivalence" which will be used later, see ibid., pp. 25-41, 70-73.) Now it happens that, for separable spaces S, there exists another proof which does not make use of so much material. Therefore two proofs will be given: first, for separable spaces S (Satisfying Postulate D_1) without the use of the general set-theory, and second, the set-theoretical proof for an arbitrary space S.

THEOREM 12.18: *There exists a complete o.n. set A in S if S satisfies Postulate D_1 (that is, if S is separable) and A is at most countable.*

Proof: Let f_1, f_2, ... be a sequence of elements of S which is dense in S. Let f_{n_1} be the first element $\neq 0$; in general, let f_{n_i} be the first element of this sequence after $f_{n_{i-1}}$ which is linearly independent of $f_{n_1}, \dots, f_{n_{i-1}}$. If for some value k of i there exists no such element f_{n_k} (k being the smallest such value of i), then the induction stops at $f_{n_{k-1}}$. Let all such elements f_{n_i} be selected. An o.n. set φ_1, φ_2, ... may be constructed from the elements f_{n_i} by the process used in the proof of Theorem 12.13. It follows that every f_{n_i} is a linear aggregate of φ's and conversely. And as every f_μ is an f_{n_i} or a linear aggregate of f_{n_i}'s, every f_μ is a linear aggregate of φ's.

The φ's form an o.n. set A by construction. If f is orthogonal to all the φ's, it is orthogonal to their linear aggregates and therefore to all the f_μ's and the condensation points of the f_μ's, that is to say, to every element g of S. Thus f is orthogonal to itself and f = 0. Hence A is a complete o.n. set.

THEOREM 12.19. *There exists a complete o.n. set A in S.*

Proof: Let the elements of S be well ordered so that to each element of S there is attached an ordinal number α. The elements φ_α of A will be selected from among the elements f_α of S by transfinite induction. If all φ_β, $\beta < \alpha$, have already been selected, then proceed for α as follows: if $\|f_\alpha\| = 1$ and $(f_\alpha, \varphi_\beta) = 0$ for every $\beta < \alpha$ for which a φ_β has been defined, then put $\varphi_\alpha = f_\alpha$; if f_α does not fulfill these conditions, then leave φ_α undefined. It is obvious that the set A so determined is o.n. Suppose there existed an element f of S, $\|f\| = 1$, orthogonal to A. This f has an ordinal number α, $f = f_\alpha$, and it would be the case that $\|f_\alpha\| = 1$ and $(f_\alpha, \varphi_\beta) = 0$ for all $\beta < \alpha$. Hence $f = f_\alpha = \varphi_\alpha$ would have been included in A. Therefore A is complete.

It is of interest to inquire which subsets M of S satisfy Postulates A, B, etc. (of course when the definitions of f + g, af, and (f, g) remain unchanged). One sees immdiately that the answer is as follows: M satisfies A (linearity) if and only if M is linear. M always satisfies B (existence of an inner product). M satisfies E_1 (completeness) if and only if it is closed.

Thus the c.l.m.'s are those subsets of S which satisfy A, B, and E_1, the postulates which have been explicitly assumed to hold for S. Therefore the c.l.m.'s are of particular importance.

With regard to the other postulates, the following remarks are apparent : if S satisfies C_1, then any subset M does also, and $N_2 \leqq N_1$, where N_1 is the N associated with S and N_2 that associated with M. If S satisfies C_2, M may satisfy either C_1 or C_2. If S satisfies D_1, M does also (Theorem 12.12), and if S satisfies D_2, M may satisfy either D_1 or D_2.

THEOREM 12.20. *If M is a c.l.m., a necessary and sufficient condition that an o.n. set $A \subset M$ be complete in M is that any element f of M be representable as* $\sum_{\varphi \in A} (f, \varphi)\varphi$.

Proof: The sufficiency of the condition is evident, for if $f \in M$ and $f \perp A$, $f = 0$. The condition is necessary, for if f is any element of M, then $\sum_{i=1}^{n} (f, \varphi_i)\varphi_i$ is an element of M (where $\varphi_1, \ldots, \varphi_n$ are elements of A), and $\sum_{\varphi \in A} (f, \varphi)\varphi$ is convergent and is an element of M since M is closed. Hence $f - \sum_{\varphi \in A} (f, \varphi)\varphi$ is an element of M which, by Theorem 12.17, is orthogonal to A. Since A is complete, $f - \sum_{\varphi \in A} (f, \varphi)\varphi = 0$ and $f = \sum_{\varphi \in A} (f, \varphi)\varphi$.

THEOREM 12.21: *If M is a c.l.m., a necessary and sufficient condition that an o.n. set A in M be complete in M is that, for any element f of M,* $\|f\|^2 = \sum_{\varphi \in A} |(f, \varphi)|^2$.

Proof: The condition is sufficient, for if f is an element of M orthogonal to A, then $\|f\| = 0$ and $f = 0$. The condition is necessary, for, by Theorem 12.20, any element f of M may be represented as $f = \sum_{\varphi \in A} (f, \varphi)\varphi$. Hence $(f, f) = \|f\|^2 = ([\sum_{\varphi \in A} (f, \varphi)\varphi], f) = \sum_{\varphi \in A} (f, \varphi)(\varphi, f) = \sum_{\varphi \in A} |(f, \varphi)|^2$.

Corollary: *In the preceding theorem the condition* $\|f\|^2 = \sum_{\varphi \in A} |(f, \varphi)|^2$ *is equivalent to the condition that* $(f, g) = \sum_{\varphi \in A} (f, \varphi)(\varphi, g)$ *for any two elements f and g of M.* ($\sum_{\varphi \in A}$ *converges absolutely.*)

Proof: The second condition obviously implies the first. Conversely, if in the first condition f is replaced by f + g and f - g and the two results subtracted, the condition $\mathfrak{R}(f, g) = \sum_{\varphi \in A} \mathfrak{R}[(f, \varphi)(\varphi, g)]$, as well as the absolute convergence of $\sum_{\varphi \in A} \mathfrak{R}$, result. If herein f and g are replaced by if and g, then it is seen that $\mathfrak{I}(f, g) = \sum_{\varphi \in A} \mathfrak{I}[(f, \varphi)(\varphi, g)]$, and the absolute convergence of $\sum_{\varphi \in A} \mathfrak{I}$ follows. Hence $(f, g) = \sum_{\varphi \in A} (f, \varphi)(\varphi, g)$.

Definition 12.14: *If M is a subset of S, then* $\ominus M$ *is the set of all elements of S orthogonal to M.*

THEOREM 12.22: $M \subset N$ *implies that* $\ominus M \supset \ominus N$. $\ominus M$ *is always a c.l.m.* $\ominus M = \ominus \{M\} = \ominus [M]$. $\ominus [M, N, \ldots] = (\ominus M) \cdot (\ominus N) \cdot \ldots$.

Proof: That $M \subset N$ implies that $\ominus M \supset \ominus N$ is clear.

The set Q_f of all elements g of S orthogonal to a given element f is obviously a c.l.m. Since $\ominus M = \prod_{f \in M} Q_f$, it follows that $\ominus M$ is a c.l.m. (Note the second remark after Definition 12.13.)

Suppose $f \in \ominus M$. Then $M \perp f$ and $Q_f \supset M$. Since Q_f is a c.l.m., $Q_f \supset [M]$. Hence $[M] \perp f$, that is, $f \in \ominus[M]$. Therefore $\ominus M \subset \ominus [M]$. By the first part of the theorem, $\ominus M \supset \ominus \{M\} \supset \ominus [M]$ since $M \subset \{M\} \subset [M]$. Hence

$\ominus M = \ominus\{M\} = \ominus[M]$.

If $f \in (\ominus M) \cdot (\ominus N) \cdot \ldots$, then $f \in \ominus M$, $f \in \ominus N$, ..., so that $f \perp M$, $f \perp N$, ..., and therefore $f \perp P$, where P is the sum of M, N, ... Conversely, if $f \perp P$, then $f \in (\ominus M) \cdot (\ominus N) \cdot \ldots = \ominus P = \ominus[P] = \ominus[M, N, \ldots]$. Thus all parts of the theorem are proved.

<u>THEOREM 12.23</u>: <u>If</u> M <u>is</u> <u>a</u> <u>c.l.m.</u> <u>in</u> S, <u>every element</u> f <u>of</u> S <u>may be represented in one and only one way as</u> $f = f_1 + f_2$, <u>where</u> $f_1 \in M$ <u>and</u> $f_2 \in \ominus M$.

Proof: By applying Theorem 12.19 to the c.l.m. M (instead of S), it follows that there exists an o.n. set A in M which is complete in M. Let $f_1 = \sum_{\varphi \in A} (f, \varphi)\varphi$ and let $f_2 = f - f_1$. As in the proof of Theorem 12.20, $f_1 \in M$ and $f_2 \perp A$. Hence $f_2 \in \ominus A$ and $f_2 \in \ominus[A]$. By Theorem 12.20, $[A] = M$, so that $f_2 \in \ominus M$.

This representation is unique, for if $f = f_1 + f_2 = f_1' + f_2'$ (where f_1 and f_1' are in M and f_2 and f_2' are in $\ominus M$, then $f_1 - f_1' = f_2' - f_2$. But $f_1 - f_1'$ is in M and $f_2' - f_2$ is in $\ominus M$. Hence the element represented by these two expressions is orthogonal to itself so that it must be 0. Thus $f_1 = f_1'$ and $f_2 = f_2'$.

<u>THEOREM 12.24</u>: $\ominus(\ominus M) = [M]$, <u>or, if</u> M <u>is a c.l.m.</u>, $\ominus(\ominus M) = M$.

Proof: As $\ominus(\ominus M) = \ominus(\ominus[M])$ by Theorem 12.23, and as [M] is a c.l.m., the first part of the theorem follows from the second if M is replaced by [M]. So it is necessary to consider only the second part. As every element of M is orthogonal to $\ominus M$, $M \subset \ominus(\ominus M)$. Thus it remains to show that $\ominus(\ominus M) \subset M$.

Suppose $f \in \ominus(\ominus M)$. By Theorem 12.23, $f = f_1 + f_2$, where $f_1 \in M$ and $f_2 \in \ominus M$. By the preceding argument, $f_1 \in \ominus(\ominus M)$, so that $f - f_1 = f_2$ is in $\ominus(\ominus M)$. Since f_2 is in both $\ominus M$ and $\ominus(\ominus M)$, it is orthogonal to itself and must be 0. Hence $f = f_1$ and $f \in M$. Thus $\ominus(\ominus M) \subset M$, and the proof is complete.

It follows from Theorems 12.18 and 12.20 that in every space S satisfying Postulates C_2 and D_1 (that is, separable but not Euclidean), there exists a complete o.n. set A: $\varphi_1, \varphi_2, \ldots$ in S, which must be countably infinite, such that each element f of S can be represented as $f = \sum_{i=1}^{\infty} (f, \varphi_i)\varphi_i = \sum_{i=1}^{\infty} x_i \varphi_i$, where $x_i = (f, \varphi_i)$. Hence to each element f there corresponds one and only one set of complex numbers $(x_1, x_2, \ldots)$ such that, by Corollary 1 of Theorem 12.11, $\sum_{i=1}^{\infty} |x_i|^2$ is finite. Conversely, to each set of complex numbers$(x_1, x_2, \ldots)$ such that $\sum_{i=1}^{\infty} |x_i|^2$ is finite, there corresponds, by Theorem 12.16, one and only one element f of S: $f = \sum_{i=1}^{\infty} x_i \varphi_i$.

If f corresponds to $(x_1, x_2, \ldots)$ and g to $(y_1, y_2, \ldots)$, then af obviously corresponds to $(ax_1, ax_2, \ldots)$ and f + g to $(x_1 + y_1, x_2 + y_2, \ldots)$, and, by the Corollary to Theorem 12.21, $(f, g) = \sum_{i=1}^{\infty} x_i\bar{y}_i$.

Thus S (which was assumed to satisfy Postulate D_1) is isomorphic with the space given by

<u>Definition 12.15: The space of all sequences of complex numbers</u> $(x_1, x_2, \ldots)$ <u>such that</u> $\sum_{i=1}^{\infty} |x_i|^2$ <u>is finite and in which the operations</u> af, f + g, (f, g) <u>are defined by</u>

$$a(x_1, x_2, \ldots) = (ax_1, ax_2, \ldots),$$

$$(x_1, x_2, \ldots) + (y_1, y_2, \ldots) = (x_1 + y_1, x_2 + y_2, \ldots),$$

$$((x_1, x_2, \ldots), (y_1, Y_2, \ldots)) = \sum_{i=1}^{\infty} x_i\bar{y}_i$$

<u>is called Hilbert space</u>, H.

It will be shown that af and f + g belong to H, and that the summation defining (f, g) converges absolutely. If a space S satisfying Postulates A, B, C_2, D_1, and E_1 were now known, these assertions would follow from the isomorphism of S and H deduced above, as would also the fact that H too satisfies these postulates. But since no such space has been constructed here (some will be given in the appendix to this chapter; H is the simplest), these assertions must be proved independently.

The absolute convergence of $\sum_{i=1}^{\infty} x_i\bar{y}_i$ follows from that of $\sum_{i=1}^{\infty} |x_i|^2$ and $\sum_{i=1}^{\infty} |y_i|^2$ together with the fact that $|x_i\bar{y}_i| = |x_i| \cdot |y_i| \leqq \frac{1}{2}|x_i|^2 + \frac{1}{2}|y_i|^2$. The remaining assertions are provided for by

THEOREM 12.25: H satisfies Postulates A, B, C_2, D_1, and E_1.

Proof: The relations a) to g) of A (Definition 12.1) are obvious, it being necessary to prove only that if f and g belong to H, then af and f + g do also. But if $\sum_{i=1}^{\infty}|x_i|^2$ is finite, then $\sum_{i=1}^{\infty} |ax_i|^2$ is also finite. Again, if $\sum_{i=1}^{\infty} |x_i|^2$ and $\sum_{i=1}^{\infty} |y_i|^2$ are finite, then $\sum_{i=1}^{\infty}|x_i + y_i|^2$ is finite since $|x_i + y_i|^2 + |x_i - y_i|^2 = 2|x_i|^2 + 2|y_i|^2$, so that $|x_i + y_i|^2 \leqq 2|x_i|^2 + 2|y_i|^2$.

It is apparent that B is satisfied.

Since the elements $f_j = (0, \ldots, 0, 1, 0, \ldots)$ (the 1 is at the jth place), $j = 1, \ldots, n$, are linearly independent, C_2 is satisfied.

That H satisfies D_1 (separability) follows from the fact that the set of all elements $(\xi_1, \ldots, \xi_n, 0, 0, \ldots)$, $n = 1, 2, \ldots$; $\mathfrak{R}(\xi_1), \mathfrak{I}(\xi_1), \ldots, \mathfrak{R}(\xi_n), \mathfrak{I}(\xi_n)$ rational, is countable and dense in H. (If $(x_1, x_2, \ldots)$ is in H, there exists an n and an element $(\xi_1, \ldots, \xi_n, 0, 0, \ldots)$ such that, corresponding to $\varepsilon > 0$, $\sum_{i=n+1}^{\infty} |x_i|^2 < \frac{\varepsilon}{2}$ and $|\xi_i - x_i|^2 < \frac{\varepsilon}{2n}$, $i = 1, \ldots, n$.)

Finally, consider the fundamental sequence of elements $f_n = (x_1^{(n)}, x_2^{(n)}, \ldots)$, $n = 1, 2, \ldots$. Since $\lim_{m,n\to\infty} \|f_m - f_n\| = 0$, $\lim_{m,n\to\infty} \sum_{i=1}^{\infty}|x_i^{(m)} - x_i^{(n)}|^2 = 0$, so that for each i, $\lim_{m,n\to\infty} |x_i^{(m)} - x_i^{(n)}| = 0$. Thus $x_i = \lim_{n\to\infty} x_i^{(n)}$ exists. But $\lim_{m\to\infty} \left(\lim_{n\to\infty} \sum_{i=1}^{\infty} |x_i^{(m)} - x_i^{(n)}|^2 \right) = 0$, so that $\lim_{m\to\infty} \sum_{i=1}^{\infty} |x_i^{(m)} - x_i|^2 = 0$. Thus at least one of the numbers $\sum_{i=1}^{\infty} |x_i^{(m)} - x_i|$ is finite (say for $m = m_0$), $(x_1^{(m_0)} - x_1, x_2^{(m_0)} - x_2, \ldots)$ is in H, and, as $(x_1^{(m_0)}, x_2^{(m_0)}, \ldots)$ is in H,

$f = (x_1, x_2, \ldots)$ is in H. Since $\lim_{m\to\infty} \sum_{i=1}^{\infty} |x_i^{(m)} - x_i|^2 = 0$, $\lim_{m\to\infty} f_n = f$. Hence E_1 is satisfied.

The entire discussion from Theorem 12.24 on shows that there is one and, up to isomorphisms, essentially only one space S satisfying A, B, C_2, D_1, and E_1, i.e., the Hilbert space H.

Now consider the case where S satisfies D_2.

Again there is a complete o.n. set A in S, but it is not countable. A may be mapped in a one-to-one fashion on some set I of indices. For example, if A were finite, I might be chosen as the set of positive integers 1, ..., N (where N is the number of elements of A); if A were countably infinite, I might be chosen as the set of all positive integers 1,2,...; in any event, because of the well ordering theorem, I may be chosen as the set of all ordinal numbers $\alpha < \Omega$, where Ω is a suitably chosen "aleph" (i.e., that for I; for these notions, cf. loc.cit. before Theorem 12.18). Of course, the map may be chosen in other ways, and it might even coincide with A.

If $\alpha \in I$, denote the element of A corresponding to α in the given one-to-one mapping of A on I by φ_α.

By Theorem 12.20, each element f of S can be represented as $f = \sum_{\alpha\in I} (f, \varphi_\alpha)\varphi_\alpha = \sum_{\alpha\in I} x_\alpha \varphi_\alpha$, where $x_\alpha = (f, \varphi_\alpha)$. Hence to each element f there corresponds one and only one ordered set of complex numbers (x_α), $\alpha \in I$, such that, by Corollary 2 of Theorem 12.11, $x_\alpha \neq 0$ for only a countable set of α's and $\sum_{\alpha\in I} |x_\alpha|^2$ is finite. Conversely, to each ordered set of complex numbers (x_α), $\alpha \in I$, where $x_\alpha \neq 0$ for only a countable set of α's and $\sum_{\alpha\in I} |x_\alpha|^2$ is finite, there corresponds, by Theorem 12.16, one and only one element f of S: $f = \sum_{\alpha\in I} x_\alpha \varphi_\alpha$.

If f corresponds to (x_α) and g to (y_α), then af corresponds to (ax_α) and f + g to $(x_\alpha + y_\alpha)$, and, by the Corollary to Theorem 12.21, (f, g) =

$= \sum_{\alpha \in I} x_\alpha \bar{y}_\alpha$.

Thus S (which was assumed to satisfy D_2) is isomorphic with the space given by the following definition (in which it is not assumed that I is non-countable or even that I is infinite; but the I corresponding to the space S discussed above will of course be non-countable).

Definition 12.16: Let I be an arbitrary set (of indices α). The space of all ordered sets of complex numbers (x_α), $\alpha \in I$, such that $x_\alpha \neq 0$ for only a countable set of α's (if I itself is countable, this condition is omitted) and $\sum_{\alpha \in I} |x_\alpha|^2$ is finite, and in which the operations af, f + g, and (f,g) are defined by

$$a(x_\alpha) = (ax_\alpha),$$
$$(x_\alpha) + (y_\alpha) = (x_\alpha + y_\alpha),$$
$$((x_\alpha), (y_\alpha)) = \sum_{\alpha \in I} x_\alpha \bar{y}_\alpha$$

is called H_I.

The sum $\sum_{\alpha \in I} x_\alpha \bar{y}_\alpha$ is absolutely convergent, again because $|x_\alpha \bar{y}_\alpha| = |x_\alpha| \cdot |y_\alpha| \leq \frac{1}{2}|x_\alpha|^2 + \frac{1}{2}|y_\alpha|^2$. In other respects, H_I is characterized by

THEOREM 12.26: If I is finite with N elements, H_I is isomorphic with the N-dimensional complex Euclidean space, that is, H_I satisfies A, B, C_1, D_1, and E_1. (For I =(1, ..., N), H_I is identical with it.) If I is countably infinite, H_I is isomorphic with Hilbert space H, that is H_I satisfies A, B, C_2, D_1, and E_1. (For I = (1, 2, ...), H_I is identical with it.) If I is non-countable H_I satisfies A, B, C_2, D_2, and E_1.

Proof: The first two statements are obvious. (See Theorems 12.14 and 12.25.) The third is proved for A, B, C_2, and E_1 in the same way as in Theorem 12.25. D_2 follows, by Corollary 5 of Theorem 12.11, from the existence of a non-countable o.n. set, i.e., the set of all elements $\varphi_\beta = (\delta_{\alpha\beta})$, where α, $\beta \in I$,

$$\delta_{\alpha\beta} = \begin{cases} 1 & \text{if } \alpha = \beta, \\ 0 & \text{if } \alpha \neq \beta. \end{cases}$$

It is now possible to characterize completely the space S satisfying A, B, and E_1.

THEOREM 12.27: Every space S satisfying A, B, and E is isomorphic with some space H_I. With regard to isomorphism, I may be replaced by any set I' of the same "power" (that is, any set I' on which I can be mapped in a one-to-one fashion). But every change of "power" destroys the isomorphism, that is if I and I' are of different powers, H_I and $H_{I'}$ are non-isomorphic. Thus if I is to be chosen as the set of all ordinal numbers $\alpha < \Omega$, where Ω is any aleph (cf. loc..cit.before Theorem 12.18), then Ω is determined uniquely by S.

In the sense of Theorem 12.26, the relation between C_ρ, D_σ, and I or Ω is as follows: if I is finite, so that $\Omega = N = 1, 2, \ldots < \omega$, then S satisfies C_1 and D_1; if I is countably infinite, so that $\Omega = \omega$, then S satisfies C_2 and D_1; if I is non-countable, so that $\Omega > \omega$, then S satisfies C_2 and D_2.

Proof: The first statement follows from Theorem 12.14 and the remarks before Definitions 12.15 and 12.16. The second statement is obvious. The fourth follows from the third. The last statement is a repetition of Theorem 12.26. It remains to prove the third statement, i.e., if H_I and $H_{I'}$ are isomorphic, I can be mapped in a one-to-one fashion on I'.

Now if I is finite, H_I is a complex Euclidean space of (say) N dimensions (Theorem 12.14) and (by the remark on complete o.n. sets following this theorem) I and I' must both have exactly N elements. Thus the third statement is proved if I is finite, and similarly if I' is finite, so that it may be assumed that both I and I' are infinite.

Defino $\varphi_\beta = (\delta_{\alpha\beta})$, where $\alpha, \beta \in I$, $\delta_{\alpha\beta} = \begin{cases} 1 & \text{if } \alpha = \beta, \\ 0 & \text{if } \alpha \neq \beta, \end{cases}$ and let the element of $H_{I'}$ corresponding to φ_β in the isomorphism be φ'_β. In $H_{I'}$, $\varphi'_\beta = (x_{\alpha'\beta})$, $\alpha' \in I'$, $\beta \in I$. For a given $\beta \in I$, denote the countable set of α''s ($\in I'$) for which $x_{\alpha'\beta} \neq 0$ by Y_β.

If $\alpha'_0 \in I'$ belongs to no Y_β, then $\psi' = (\delta_{\alpha'\alpha'_0})$, $\alpha' \in I'$, is normalized and orthogonal to every $\varphi'_{\alpha'}$, so that its image $\psi = (y_\alpha)$, $\alpha \in I$, in I by the isomorphism is normalized and orthogonal to every φ_β. Since the condition $(\psi, \varphi_\beta) = 0$ implies that $y_\beta = 0$, therefore $\psi = 0$. This contradicts the fact that ψ is normalized.

Therefore I' is exhausted by the sum of all the Y_β's, so that its "power" is not greater than $\aleph_0$ times the "power" of I. But infinite powers are unaltered when multiplied by $\aleph$ (cf.loc.cit., p.71). Hence the power of I' is not greater than the power of I. Interchange of I and I' shows that the power of I is not greater than the power of I'. Hence I and I' have the same power. This completes the proof of all parts of the theorem.

This characterization of the spaces S satisfying A, B, and E_1 leads to exactly one invariant, i.e., the aleph Ω.

<u>Definition 12.17: The aleph Ω, which is an isomorphism-invariant uniquely determined by S and which completely characterizes S to within isomorphisms (Theorem 12.27), is called the dimension of S.</u>

The spaces S with $\Omega = N = 1, 2, \ldots < \omega$, $H_{(1,\ldots,N)}$ are complex Euclidean spaces of finite dimension and lead to nothing new. But $\Omega = \omega$ leads to the Hilbert space, $H = H_{(1,2,\ldots)}$, which is a new and interesting geometrical object. The cases $\Omega > \omega$ correspond to still more general spaces, $H_{(\alpha)\alpha<\Omega}$ but it will be seen that these spaces are very similar to Hilbert space with regard to most of their important properties. This is due to the fact that every separable part of them lies in a subset of them which is a

Hilbert space, that is, in a separable c.l.m. Thus all theorems which do not refer simultaneously to more than a countable or separable set of points are really always discussed in Hilbert space.

In order to obtain this result, the following theorem is needed:

THEOREM 12.28: *If X is a subset of S, then either all or none of the sets X, {X}, [X] are separable.*

Proof: If {X} is separable, then [X] is also, since {X} is dense in [X]. If [X] is separable, then X is also, since $X \subset [X]$. It remains to show that if X is separable, then {X} is also.

Suppose the sequence $f_1, f_2, \ldots$ is dense in X. If $\mathfrak{R}(a_1^\circ), \mathfrak{I}(a_1^\circ), \ldots, \mathfrak{R}(a_n^\circ), \mathfrak{I}(a_n^\circ)$, $n = 1, 2, \ldots$, are all rational, then the elements $a_1^\circ f_1 + \ldots + a_n^\circ f_n$ form a countable set D which is dense in the set of all elements $a_1 f_1 + \ldots + a_n f_n$, where $n = 1, 2, \ldots$ and where $a_1, \ldots, a_n$ are arbitrary complex numbers. Thus D is dense in the set of all elements $a_1 g_1 + \ldots + a_n g_n$, where $g_1, \ldots, g_n$ are arbitrary elements of X; therefore D is dense in {X}. Hence {X} is separable, and the proof is complete.

Thus every separable X is contained in [X] which, being a separable c.l.m., is a complex Euclidean space or a Hilbert space.

Appendix I.

It is desirable to give further examples of spaces having the properties A, B, E, and some combination of C_ρ, D_σ $(\rho, \sigma = 1,2)$. In particular those with $C_2 D_1$ are of interest, as they are isomorphic with Hilbert space (cf. Theorems 12.25 or 12.26). This is particularly important, because these examples will be "functional spaces" which, on the one hand, will later be useful in illustrating many results of the theory of Hilbert spaces; and on the other hand they lead to the chief applications of the theory. Other examples, based on uncountably infinite direct products, will furnish interesting examples of "functional" hyper-Hilbert spaces.

Before we construct our examples, it is advisable to give the following extension of the definitions of Chapter XII:

THEOREM 12'.1. *Assume that a space S satisfies the postulates* A, B *of Chapter XII, except* B, b) (cf. Definition 12.3) *instead of which only this is required:*

b') $$(f, f) \geqq 0.$$

Then S *can be transformed in the following way into a space* $\bar{S}$ *which satisfies* A, B *without exception:*

Call two elements f, g *of* S *equivalent,* $f \sim g$, *if* $\|f - g\| = 0$. *Then the following facts hold:*

(i) $f \sim f$; $f \sim g$ *implies* $g \sim f$; $f \sim g$, $g \sim h$ *implies* $f \sim h$.

(ii) *Call the set of all* $g \sim f$ *for a given* $f \in S$, $\emptyset_f$. *All* $\emptyset_f$ *are mutually exclusive subsets of* S *(that is:* $\emptyset_f \neq \emptyset_g$ *implies* $\emptyset_f \cdot \emptyset_g = 0$), *and their sum is* S. *Call their set* $\bar{S}$.

(iii) $f \sim g$ *implies* $af \sim ag$; $f \sim g$, $h \sim k$ *implies* $f + h \sim g + k$; $f \sim g$, $h \sim k$ *implies* $(f, h) = (g, k)$, *and thus* $\|f\| = \|g\|$.

(iv) *If* $\emptyset, \psi \in \bar{S}$, $f \in \emptyset$, $g \in \psi$, *then* $\emptyset_{af}$, $\emptyset_{f+g}$, (f, g), $\|f\|$ *depend on* a, $\emptyset$, ψ *alone (and not on the particular choice of* $f \in \emptyset$, $g \in \psi$). *Call them* $a\emptyset$, $\emptyset + \psi$, $(\emptyset, \psi)$, $\|\emptyset\|$.

(v) *With the notations of* (iv) $\bar{S}$ *satisfies the postulates* A, B *without exception* .

(vi) $\bar{S}$ *satisfies* E *if and only if* S *satisfies it.*

(vii) $\bar{S}$ *has o.n. subsets and complete o.n. subsets of the same powers as* S.

Proof: One verifies immediately that the proofs of Theorems 12.5 (Schwarz's Lemma) and 12.6 depended on B, a), b'), c), d) alone (cf. Definition 12.3). Therefore they apply to our present S. Thus $\|f\| = 0$, $\|g\| = 0$ imply $\|af\| = 0$, $\|f + g\| = 0$. Considering $f - f = 0$, $g - f = -(f - g)$, $f - h = (f - g) + (g - h)$ this proves (i); considering $a(f - g) = af - ag$, $((f + h) - (g + k)) = (f - g) + (h - k)$ it proves the two first statements of (iii). As to the third statement of (iii), owing to the distributivity of (g, k) (cf. B, a), c), d)) we need only to consider the case where $f = 0$, $h = 0$. Then $|(g, k)| \leqq \|g\| \cdot \|k\| = 0$, $(g, k) = 0$, completing the proof.

(ii), (iv) are immediate consequences of (i), (iii) respectively.

All parts of **A** as well as B, a), b'), c),d) carry over immediately from S to $\bar{S}$. $\|\phi\| = 0$ means $\phi = \phi_f$ with $\|f\| = 0$, that is $f \sim 0$, that is $\phi = \phi_0$; and ϕ_0 plays the role of 0 in $\bar{S}$. Thus (v) is proved.

As to (vi) and (vii), it is obvious that the behavior of S and $\bar{S}$ is identical in all these respects: we pass from S to $\bar{S}$ by replacing each $f \in S$ by its ϕ_f; and from $\bar{S}$ to S by replacing each $\phi \in \bar{S}$ by some $f \in \phi$ (that is $\phi = \phi_f$).

Definition 12'1: If S satisfies the assumptions of the preceding theorem, then the formation of the $\phi = \phi_f$'s is called the process of necessary identifications in S; and $\bar{S}$ is said to arise from S by making the necessary identifications in S.

In what follows we will, for the sake of brevity, identify $S_{D,\mu^*(M)}$ and $\bar{S}_{D,\mu^*(M)}$ and correspondingly f and ϕ_f , whenever this can be done without a loss of clarity.

Let now D be a space and $\mu^*(M)$ a regular outer measure in it (defined for all subsets M of D, cf. Definitions 10.2.1, 10.2.5).

Of course, we could just as well begin with a finite, non-negative, totally additive measure function $\nu(M)$, defined on a half-ring $\mathcal{R}$ of subsets M of D. Then we would form, by Theorem 10.3.2, the outer measure $\nu_1^*(M)$, which agrees with $\nu(M)$ on $\mathcal{R}$ and is determined by $\mathcal{R}$ -- and put $\mu^*(M) = \nu_1^*(M)$. As we will see below, it would even suffice to extend $\nu(M)$ to the Borel-ring of $\mathcal{R}$, BR($\mathcal{R}$). But we prefer to use consistently the outer measure. Therefore we return to the $\mu^*(M)$ as mentioned above.

We now define:

Definition 12'2. Let D be a space, and $\mu^*(M)$ be a regular outer measure in it. Let $MF_{D,\mu^*(M)}$ be the set of all complex valued functions $f \equiv f(P)$, defined for all $P \in D$, and measurable with respect to $\mu^*(M)$.

Let $S_{D,\mu^*(M)}$ be the set of all those $f \in MF_{D,\mu^*(M)}$ for which $\int_D |f(P)|^2 \, d\mu(M_P)$ is finite.

THEOREM 12'2: We have:

(i) If $f \in S_{D,\mu^*(M)}$ and a is a complex constant, then $af \in S_{D,\mu^*(M)}$ $((af)(P) \equiv a \cdot f(P))$.

(ii) If $f, g \in S_{D,\mu^*(M)}$, then $f + g \in S_{D,\mu^*(M)}$ $((f+g)(P) \equiv f(P) + g(P))$.

(iii) If $f, g \in S_{D,\mu^*(M)}$, then $\int_D f(P)\overline{g(P)} \, d\mu(M_P)$ is finite.

(iv) If we define in $S_{D,\mu^*(M)}$ af, f + g as in (i), (ii) and $(f, g) =$ $= \int_D f(P)\overline{g(P)} d\mu(M_P)$ (cf. (iii)), the situation of Theorem 12'.1 arises: $S_{D,\mu^*(M)}$ satisfies the postulates A, B, E of Chapter XII, where B, b), must be replaced by B, b').

(v) The necessary identifications (cf. Definition 12'1) are these: $f \sim g$ if and only if the set of all P with $f(P) \neq g(P)$ has the measure 0. The $\bar{S}_{D,\mu^*(M)}$ which results from these identifications satisfies A, B, E without exception. (Cf. Theorem 12'.1 and Definition 12'.1.)

Proof: (i) is obvious; (ii) follows owing to the identity $|f(P) + g(P)|^2 = |f(P)|^2 + |g(P)|^2 + 2\,\mathcal{R}(f(P)\overline{g(P)})$, from (iii); (iii) follows from $|f(P)\overline{g(P)}| \leqq \frac{1}{2}|f(P)|^2 + \frac{1}{2}|g(P)|^2$. The statements of (iv) concerning A, B are obvious. $\|f - g\| = 0$ means $(f - g, f - g) = 0$, $\int_D |f(P) - g(P)|^2 d\mu(M_P) = 0$; and as $|f(P) - g(P)|^2 \geqq 0$ everywhere, this means that $|f(P) - g(P)|^2 \neq 0$, $f(P) \neq g(P)$ holds only on a P-set of measure 0. This proves the first half of (v), the second half follows from (iv) and Theorem 12'.1, (v), (vi). Thus it remains only to prove the statement of (iv) concerning E_1: That $S_{D,\mu^*(M)}$ is complete.

Consider a sequence $f_1, f_2, \ldots \varepsilon S_{D,\mu^*(M)}$ for which $\lim_{m,n\to\infty} \|f_m - f_n\| = 0$. Thus for every $\varepsilon > 0$ an $n_o = n_o(\varepsilon)$ exists such that $m, n \geqq n_o$ imply $\|f_m - f_n\| \leqq \varepsilon$.

Choose a subsequence $\ell_1, \ell_2, \ldots$ of $1, 2, \ldots$ $(\ell_1 < \ell_2 < \ldots)$ with $\ell_\sigma \geqq n_o\left(\frac{1}{4^\sigma}\right)$. Then $\ell_\sigma, \ell_{\sigma+1} \geqq n_o\left(\frac{1}{4^\sigma}\right)$, $\|f_{\ell_\sigma} - f_{\ell_{\sigma+1}}\| \leqq \frac{1}{4^\sigma}$,

$\int_D |f_{\ell_\sigma}(P) - f_{\ell_{\sigma+1}}(P)|^2 \, d\mu(M_P) = \|f_{\ell_\sigma} - f_{\ell_{\sigma+1}}\|^2 \leqq \frac{1}{16^\sigma}$. Denote by T_σ the set of all P's for which $|f_{\ell_\sigma}(P) - f_{\ell_{\sigma+1}}(P)| \geqq \frac{1}{2^\sigma}$; then T_σ is measurable, and the above integral is clearly $\geqq \left(\frac{1}{2^\sigma}\right)^2 \mu(T_\sigma) = \frac{\mu(T_\sigma)}{4^\sigma}$; thus $\mu(T_\sigma) \leqq \frac{1}{4^\sigma}$. Therefore

$$\mu\left(\sum_{\sigma=\rho}^{\infty} T_\sigma\right) \leqq \sum_{\sigma=\rho}^{\infty} \mu(T_\sigma) \leqq \sum_{\sigma=\rho}^{\infty} \frac{1}{4^\sigma} \leqq \frac{1}{3\cdot 4^{\rho-1}}$$

and hence $\mu\left(\prod_{\rho=1}^{\infty} \sum_{\sigma=\rho}^{\infty} T_\sigma\right) \leqq \frac{1}{3\cdot 4^{\rho-1}}$, for every $\rho = 1, 2, \ldots$, and is thus $= 0$. If $P \notin \prod_{\rho=1}^{\infty} \sum_{\sigma=\rho}^{\infty} T_\sigma$, then there exists a $\rho = \rho(P) = 1, 2, \ldots$, such that for $\sigma \geqq \rho$, $P \notin T_\sigma$. That is: $|f_{\ell_\sigma}(P) - f_{\ell_{\sigma+1}}(P)| < \frac{1}{2^\sigma}$. As the series

$\frac{1}{2^\rho} + \frac{1}{2^{\rho+1}} + \ldots$ converges, this proves the existence of $\lim_{\substack{\sigma \geqq \rho \\ \sigma \to \infty}} f_{\ell_\sigma}(P)$.

As $\prod_{\rho=1}^{\infty} \sum_{\sigma=\rho}^{\infty} T_\sigma$ is a set of measure 0, we may redefine all $f_n(P)$'s in it and put them = 0 there (cf. the first part of (v)). Then $\lim f_{\ell_\sigma}(P)$ exists everywhere. Define $f(P) = \lim_{\sigma \to \infty} f_{\ell_\sigma}(P)$. $f(P)$ is clearly measurable.

Consider $|f_n(P) - f_{\ell_\sigma}(P)|^2$. If $\sigma \longrightarrow \infty$, it converges to $|f_n(P) - f(P)|^2$, and it is always $\geqq 0$. Therefore

$$\lim_{\sigma \to \infty} \inf \int_D |(f_n(P) - f_{\ell_\sigma}(P)|^2 \, d\mu(M_P) \geqq \int_D |f_n(P) - f(P)|^2 \, d\mu(M_P).$$

(Cf. Corollary 2, Theorem 11.3.7.) Now assume $n \geqq n_0(\varepsilon)$. If σ is great enough, $\ell_\sigma \geqq n_0(\varepsilon)$ also; and thus the $\int_D \ldots d\mu(M_P)$ on the left side becomes $\leqq \varepsilon^2$.

Thus the integral is finite, $f_n - f \in S_{D,\mu^*(M)}$, and with it $f = f_n - (f_n - f) \in S_{D,\mu^*(M)}$. Now the above inequality proves $\|f_n - f\| \leqq \varepsilon$ for $n \geqq n_0(\varepsilon)$, that is $\lim_{n \to \infty} f_m = f$. This completes the proof.

We wish now to discuss the behavior of $S_{D,\mu^*(M)}$ with respect to the conditions C_ρ and D_σ of Chapter XII.

<u>Definition 12'.3: A measurable subset M of D is irreducible, if every measurable subset N of M has $\mu(N) = 0$ or $\mu(M)$.</u>

<u>THEOREM 12'.3: Every set of measure 0 is irreducible; every irreducible set has a finite measure.</u>

Proof: The first statement is obvious. As to the second, observe that we can write $D = D_1 + D_2 + \ldots$, $D_1 \subset D_2 \subset \ldots$, where all D_n are measurable sets of finite measure. (Cf. Definition 10.2.6.) Thus $D_n M \subset M$, $\lim_{n \to \infty} \mu(D_n M) = \mu(M)$. But as M is irreducible, $\mu(D_n M) = 0$, or $\mu(M)$ for each n, thus the latter holds for all sufficiently great n. Therefore $\mu(M) = \mu(D_n M) \leqq \mu(D_n)$ is finite.

<u>THEOREM 12'.4. If $\mu^*(M)$ is determined by a countable half-ring $\mathcal{R}$, and every one-point-set (P), P ε D, is measurable; then every irreducible set M is either of measure 0, or else a P ε M with $0 < \mu((P)) < +\infty$ exists such that M - (P) has the measure 0.</u>

<u>Remark</u>: It is clear that any M of this structure is irreducible, and that it determines the P in question uniquely.

Proof: We must prove that if M is irreducible and $\mu(M) > 0$, then a P ε M with $\mu(M - (P)) = 0$ exists, thus that $\mu(M) = \mu((P))$, and hence that $0 < \mu((P)) < +\infty$ (the latter by Theorem 12'3).

Write D as a sequence $D_1, D_2, \ldots$. For each $n = 1, 2, \ldots$, $\mu(D_n M) = 0$ or $\mu(D_n M) = \mu(M)$; let $n_1, n_2, \ldots$ be the n's of the first kind. Then $\mu(\sum_{\sigma=1}^{\infty} D_{n_\sigma} M) = 0$ also, and thus $\sum_{\sigma=1}^{\infty} D_{n_\sigma} M \neq M$. Choose a P ε M which is $\notin \sum_{\sigma=1}^{\infty} D_{n_\sigma} M$, that is, $P \notin D_{n_\sigma}$ for $\sigma = 1, 2, \ldots$.

Consider $\mu((P))$. If it were $< \mu(M)$, we could by Theorem 10.3.2 and the remark immediately following it, find a sequence $D_{p_1}, D_{p_2}, \ldots$ with $(P) \subset \sum_{\rho=1}^{\infty} D_{p_\rho}$; $\sum_{\rho=1}^{\infty} \mu(D_{p_\rho}) < \mu(M)$. Then for some $p = p_\rho$, $P \ \varepsilon\ D_p$, $\mu(D_p M) \leq \mu(D_p) \leq \sum_{\rho=1}^{\infty} \mu(D_{p_\rho}) < \mu(M)$. Thus $\mu(D_p M) = 0$, $p = n_\sigma$ for some $\sigma = 1, 2, \ldots$, and we should have $P \ \varepsilon\ D_{n_\sigma}$, which is impossible. Thus, since $(P) \subset M$, $\mu((P)) \leq \mu(M)$, we must have $\mu((P)) = \mu(M)$, $\mu(M - (P)) = 0$. This completes the proof.

We now prove the criterion for C_ρ:

<u>THEOREM 12'.5: $\bar{S}_{D,\mu^*(M)}$ satisfies C_1 if and only if there exists a finite number of mutually exclusive irreducible subsets $M_1, \ldots, M_N$ of D, such that</u>

$$\mu(M_n) > 0 \text{ for } n = 1, \ldots, N; \quad \mu(M - \sum_{n=1}^{N} M_n) = 0.$$

If this is the case, $\bar{S}_{D,\mu^*(M)}$ has N dimensions.

If $\mu^*(M)$ is determined by a countable half-ring $\mathcal{R}$, and every one-point-set (P), $P \in D$, is measurable, then we may assume that the M_n are one-point sets: $M_n = (P_n)$.

Proof: The last statement results from the previous ones immediately, by application of Theorem 12'.4.

Consider first the necessity of our conditions. Assume that $\bar{S}_{D,\mu^*(M)}$ fulfills C_1, and has $N' = 0, 1, 2, \ldots$ dimensions. If $M_1, \ldots, M_m$ are mutually exclusive measurable subsets of $\bar{S}_{D,\mu^*(M)}$, all with measures > 0, then write again $D = D_1 + D_2 + \ldots$, $D_1 \subset D_2 \subset \ldots$, all $\mu(D_i)$ finite (cf. Definition 10.2.6). Then $\lim_{i\to\infty} \mu(D_i M_n) = \mu(M_n)$; thus there exists an i with $\mu(D_i M_n) > 0$ for all $n = 1, \ldots, m$. Define

$$f_n(P) \begin{cases} = 1 \text{ for } P \in D_i M_n \\ = 0 \text{ otherwise} \end{cases}$$

Then $f_n \in \bar{S}_{D,\mu^*(M)}$, because $\mu(D_i M_n) = \mu(D_i)$ is finite. If $\sum_{n=1}^{\infty} a_n f_n = 0$, then the P with $\sum_{n=1}^{m} a_n f_n(P) \neq 0$ form a set of measure 0; thus there exists a $P \in D_i M_n$ with $\sum_{n=1}^{m} a_n f_n(P) = 0$. This proves $a_n = 0$ for all $n = 1, \ldots, m$; thus the $f_1, \ldots, f_m$ are linearly independent. Hence $m \leqq N'$.

Therefore the possible m are bounded. Let $m = N$ be the greatest one. We see that $N \leqq N'$.

Consider the system $M_1, \ldots, M_N$. If $\mu(D - \sum_{n=1}^{N} M_n) > 0$ we could add $D - \sum_{n=1}^{N} M_n$ to it; thus $\mu(D - \sum_{n=1}^{N} M_n) = 0$. If a measurable $M' \subset M_n$ with $\mu(M') \neq 0, \mu(M_n)$ existed, we could replace M_n by M', $M_n - M'$; thus M_n is irreducible. This establishes the necessity of our conditions, together with the fact that $\bar{S}_{D,\mu^*(M)}$ has $\geqq N$ dimensions.

Consider next the sufficiency of our conditions. Assume that $M_1, \ldots, M_N$ are given, as described in the theorem. As M_n is irreducible, $\mu(M_n)$ is finite, and thus

$$f_n(P) \begin{cases} = 1 \text{ for } P \in M_n \\ = 0 \text{ otherwise} \end{cases}$$

belongs to $\bar{S}_{D,\mu^*(M)}$.

Consider now a real-valued $f \in \bar{S}_{D,\mu^*(M)}$. The set $M_n S_P[f(P) < \lambda]$ is measurable for every λ, and $\subset M_n$, thus its measure is 0 or $\mu(M)$. If we put $\lambda = -1, -2, \ldots$, it converges to 0; thus the measure is not always $\mu(M)$; if we put $\lambda = 1, 2, \ldots$, it converges to M; thus the measure is not always 0. Therefore these λ have a finite g.l.b.: λ_n. Now $\mu(M_n S_P[f(P) < \lambda_n - \frac{1}{\rho}]) = 0$, $\mu(M_n S_P[f(P) < \lambda_n + \frac{1}{\rho}]) = \mu(M)$, and therefore $\mu(M_n S_P[\lambda_n - \frac{1}{\rho} \leqq f(P) < \lambda_n + \frac{1}{\rho}]) = \mu(M)$ for $\rho = 1, 2, \ldots$. Forming the product of all these sets (for all $\rho = 1, 2, \ldots$) we obtain : $\mu(M_n S_P[f(P) = \lambda_n]) = \mu(M)$, or : $\mu(M_n S_P[f(P) \neq \lambda_n]) = 0$.

If f is complex valued, we obtain the same thing by forming λ_n for $\mathfrak{R} f(P)$ and $\mathfrak{I} f(P)$ separately, thus obtaining λ_n' and λ_n'' ; and then setting $\lambda_n = \lambda_n' + i\lambda_n''$.

Now it is clear that $f(P) \neq \sum_{M=1}^{N} \lambda_n f_n(P)$ implies

$$P \in \sum_{M=1}^{N} M_n S_P[f(P) \neq \lambda_n] + (D - \sum_{n=1}^{N} M_n).$$

This is a sum of N + 1 sets of measure 0, thus is **itself** a set of measure 0. Therefore $f \sim \sum_{M=1}^{N} \lambda_n f_n$, and in $\bar{S}_{D,\mu^*(M)}$ $f = \sum_{M=1}^{N} \lambda_n f_n$.

Thus C_1 is fulfilled, and $\bar{S}_{D,\mu^*(M)}$ has $\leqq N$ dimensions. This establishes the sufficiency of our conditions, and proves that $\bar{S}_{D,\mu^*(M)}$ has $\leqq N$

dimensions.

Therefore the proof is completed.

As to the D_σ we can only give a sufficient criterion for D_1 (its converse being thus necessary for D_2);

<u>THEOREM 12'.6:</u> $\bar{S}_{D,\mu^*(M)}$ <u>satisfies</u> D_1 <u>if</u> $\mu^*(M)$ <u>can be determined by a countable half-ring</u> $\mathcal{R}$.

Proof: We wish to show that, if a choice of $\mathcal{R}$ has been made in which $\mathcal{R}$ is countable, then $\bar{S}_{D,\mu^*(M)}$ is separable. We shall replace $\bar{S} = \bar{S}_{D,\mu^*(M)}$ by various subsets, the separability of which implies the separability of $\bar{S}$.

Consider first the set $\bar{S}_1$ of all real and non-negative functions in $\bar{S}$. If $f(P) \in \bar{S}$, then obviously $\mathcal{R} f(P)$ and $\mathcal{J} f(P) \in \bar{S}$; and thus $f_0(P) = \text{Max}\,(0, \mathcal{R} f(P))$, $f_2(P) = \text{Max}(0, -\mathcal{R} f(P))$, $f_1(P) = \text{Max}\,(0, \mathcal{J} f(P))$, $f_3(P) = \text{Max}\,(0, -\mathcal{J} f(P))$ are in $\bar{S}$ and therefore in $\bar{S}_1$. Now $f = \sum_{\rho=0}^{3} i^\rho f_\rho$. And so, if the sequence $f^{(1)}, f^{(2)}, \ldots$ is dense in $\bar{S}_1$, then the countable set of all $\sum_{\rho=0}^{3} i^\rho f^{(n_\rho)}$; $n_0, n_1, n_2, n_3 = 1, 2, \ldots$; is dense in $\bar{S}$. Thus we need only prove that $\bar{S}_1$ is separable.

Now we know that if an $f(P) \geqq 0$ is measurable, there exists a sequence of measurable finite-valued functions $f_1(P), f_2(P), \ldots$ with rational values, such that $0 \leqq f_1(P) \leqq f_2(P) \leqq \ldots \leqq f(P)$, $f_n(P) \longrightarrow f(P)$. (define $f_n(P)$ = the greatest number of the form $\frac{p}{2^n}$, $p = 0, 1, 2, \ldots$, which is $\leqq f(P)$.) Thus $\int |f(P) - f_n(P)|^2 d\mu(M_P) \longrightarrow 0$, $\|f - f_n\| \longrightarrow 0$, that is: The set $\bar{S}_2$ of all functions in $\bar{S}$, which have only a finite number of different values, all of those being rational, is dense in $\bar{S}_1$. Hence we need only prove that $\bar{S}_2$ is separable.

For any measurable subset M of D which has a finite measure, let $1_M(P)$ be defined by

$$1_M(P) \begin{cases} = 1 & \text{for } P \in M \\ = 0 & \text{for } P \notin M \end{cases}$$

Then $\bar{S}_2$ is the set of all functions of the form $\sum_{\sigma=1}^{n} \rho_\sigma 1_{M_\sigma}$, where $n=0,1,2,\ldots$; $\rho_1,\ldots,\rho_n$ rational; $M_1, \ldots, M_n$ measurable with a finite measure. Hence if $\bar{S}_3$ is the space of all functions 1_M, M as above, and if $1_{M^{(1)}}, 1_{M^{(2)}}, \ldots$ is a sequence dense in $\bar{S}_3$, then the countable set of all $\sum_{\sigma=1}^{n} \rho_\sigma 1_M$ $(n = 0,1,2,\ldots;$ $\rho_1,\ldots,\rho_n$ rational; $p_1, \ldots, p_n = 1, 2, \ldots)$ is dense in S_2. Thus we need only prove the separability of $\bar{S}_3$.

If we wish, we may consider the M's themselves instead of the 1_M's as elements of $\bar{S}_3$; only we must then define

$$\text{Distance } (M, N) = \|1_M - 1_N\| = \int |1_M(P) - 1_N(P)|^2 \, d\mu(M_P) =$$

$$= \int_{(M+N-MN)} 1 \cdot d\mu(M_P) = \mu((M + N) - MN).$$

For any of our M's we have

$$\mu(M) = \mu^*(M) = \text{l.u.b.} \sum_{i=1}^{\infty} \mu(N_i), \text{ where all } N_i \in \mathcal{R}, \text{ and } M \subset \sum_{i=1}^{\infty} N_i.$$

Thus if $\varepsilon > 0$, there exists a sequence $N_1, N_2, \ldots \in \mathcal{R}$, $M \subset \sum_{i=1}^{\infty} N_i$ with $\mu(M) + \varepsilon \geqq \sum_{i=1}^{\infty} \mu(N_i) \geqq \mu(\sum_{i=1}^{\infty} N_i)$, $\mu(\sum_{i=1}^{\infty} N_i - M) \leqq \varepsilon$. Now the sequence of sets $\sum_{i=1}^{n} N_i$, $n = 1, 2, \ldots$, converges increasingly toward $\sum_{i=1}^{\infty} N_i$; therefore $\mu(\sum_{i=1}^{n} N_i) \to \mu(\sum_{i=1}^{\infty} N_i)$, and there is an $n = 0, 1, 2, \ldots$ with $\mu(\sum_{i=1}^{n} N_i) \geqq$ $\geqq \mu(\sum_{i=1}^{\infty} N_i) - \varepsilon$, $\mu(\sum_{i=1}^{\infty} N_i) \leqq \varepsilon$. Therefore

$$\text{Distance } (M_1, \sum_{i=1}^{n} N_i) = \mu((M + \sum_{i=1}^{n} N_i) - M \sum_{i=1}^{\infty} N_i) \leqq$$

$$\leqq \mu((\sum_{i=1}^{\infty} N_i - M) + (\sum_{i=1}^{\infty} N_i - \sum_{i=1}^{n} N_i)) \leqq$$

$$\leq \mu(\sum_{i=1}^{\infty} N_i - M) + \mu(\sum_{i=1}^{\infty} N_i - \sum_{i=1}^{n} N_i) \leq 2\varepsilon \ .$$

Thus, the set $\bar{S}_4$ of all sets $\sum_{i=1}^{n} N_i$, $n = 0, 1, 2, \ldots$; $N_1, \ldots, N_n \varepsilon \mathcal{R}$; is dense in $\bar{S}_3$. But the countability of $\mathcal{R}$ obviously implies that of $\bar{S}_4$, and thus $\bar{S}_3$ is separable. It follows from our previous remarks, that this completes the proof.

The theorems which we have proved enable us to construct many examples of functional spaces which satisfy postulates A, B, E and some combination of C_ρ, D_σ, C_2, D_1, that is Hilbert space, is of chief interest here. In order to illustrate the reach and working of this method, we shall discuss some examples.

Example 1. Consider a discrete measure (cf. Chapter X, §5, example 1). In this case we know that the space D must be countable; thus it is a finite or infinite sequence of points P_1, P_2, ..., for which we may as well write 1, 2, ... Each point has a weight $w_\mu \geq 0$. The general element of $\bar{S}_{D,\mu^*(M)}$, a function f(n), n = 1, 2, ..., may be as well written as a (finite or infinite) sequence x_1, x_2, ... Then $\bar{S}_{D,\mu^*(M)}$ is defined by the requirement that

$$\sum_n w_n |x_n|^2$$

be finite -- which condition is void if the set D (of all n) is finite, and even if the set of all n with $w_n > 0$ is finite; but an essential restriction if infinitely many n's with $w_n > 0$ exist. The inner product for $f \sim (x_1, x_2, \ldots)$, $g \sim (y_1, y_2, \ldots)$ is consequently (by Theorem 12'.6),

$$(f, g) = \int x_n \bar{y}_n \, d\mu(M_n) = \sum_{n=1}^{\infty} w_n x_n \bar{y}_n \ .$$

As $\mathcal{R}$ is countable (cf. Definition 10.5.1), $\bar{S}_{D,\mu^*(M)}$ fulfills D_1. As the only irreducible sets with a measure > 0 are the one-point-sets (n) with $w_n > 0$, therefore we have C_1 or C_2 according as a finite or infinite number of n's with $w_n > 0$ exist, and their number is the dimension of $\bar{S}_{D,\mu^*(M)}$ (by Theorem 12'.5).

If, in particular all $w_n = 1$, then we are back to the normal forms of Definitions 12.15 and 12.16. But we obtain only the cases with dimensions $= 0, 1, 2, \ldots, \aleph_0$ that is the finite dimensional Euclidean spaces and Hilbert space.

Example 2. Let D be an N-dimensional Euclidean space R_N, $N = 1, 2, \ldots$, or a subset of it. We could choose for $\mu^*(M)$ the exterior measure criginating from any Lebesgue-Stieltjes-Radon-measure (cf. Chapter X, §5, Example 4). We shall only consider for the moment those $\mu(M)$'s which are absolutely continuous with respect to N-dimensional Lebesgue measure $\mu_{\mathcal{L}}(M)$, that is

$$\mu(M) = \int_M W(P)d\,\mu_{\mathcal{L}}(M_P),$$

where W(P) is any measurable, real, and non-negative function (cf. Chapter XI, §2, 3).

Now $\bar{S}_{D,\mu^*(M)}$ is the set of all measurable functions f(P) in D for which

$$\int_D |f(P)|^2 d\mu(M_P) = \int_D W(P)\ |f(P)|^2 d\mu_{\mathcal{L}}(M_P)$$

is finite. The inner product is

$$(f,\ g) = \int_D f(P)\overline{g(P)}d\mu(M_P) = \int_D W(P)\ f(P)\overline{g(P)}d\mu_{\mathcal{L}}(M_P).$$

We have $W(P) \geqq 0$ in all of D, but as the omission of all points with $W(P) = 0$ from D obviously does not matter, we may as well assume $W(P) > 0$ in all of D.

$\mathcal{R}$ can be chosen countable (cf. Theorem 10.5.19; thus $\bar{S}_{D,\mu^*(M)}$ has the property D_1 (by Theorem 12'6). As a countable $\mathcal{R}$ exists, the second half of Theorem 12'.5 shows that the M_n occurring in that theorem may be chosen as one-point-sets. This contradicts in our case $\mu(M_n) > 0$; thus $\bar{S}_{D,\mu^*(M)}$ can be in case C_1 only if $\mu(D) = 0$. As $W(P) > 0$ in all D, this means that $\mu_{\mathcal{L}}(D) = 0$. When $\bar{S}_{D,\mu^*(M)}$ has dimension 0, it contains only the element 0. Otherwise we have the case C_2; that is, the combinations C_2, D_1: then $\bar{S}_{D,\mu^*(M)}$ is a Hilbert space.

Most of the important illustrations and applications of the theory of Hilbert space are of this type.

Example 3: All foregoing examples lead to spaces fulfilling D_1, that is to separable ones. In other words: the dimension was always $\leqq \aleph_0$. We wish to show now that any infinite dimension can be obtained.

Let I be an arbitrary (infinite) set of indices, and form for each $\alpha \in I$ the same set: (0, 1), with the discrete measure $w_0 = w_1 = \frac{1}{2}$. This is the case Σ_2 in Examples 1a, 2, of §5, Chapter X. Form now the direct product $\prod\prod_{\alpha \in I} \Sigma_2$, and its measure $\tilde{\mu}(M)$, by Definition 10.4.3. Put $D = \prod\prod_{\alpha \in I} \Sigma_2$, $\mu^*(M) = \tilde{\mu}^*(M)$, and form their $\bar{S}_{D,\mu^*(M)}$. We shall see that $\bar{S}_{D,\mu^*(M)}$ has the dimension power(I). As this is $\geqq \aleph_0$, we have at any rate Case C_2; and according as it is $= \aleph_0$ or $> \aleph_0$ we shall have Case D_1 or D_2.

$D = \prod\prod_{\alpha \in I} \Sigma_2$ is the set of all functions $x = x(\alpha)$, where $\alpha \in I$, and the values are 0, 1. Define now a function $F_{\alpha_0}(x)$, for each $\alpha_0 \in I$, by

$$F_{\alpha_0}(x) = 2x(\alpha_0) - 1.$$

Clearly $F_{\alpha_0}(x)$ is always $= \pm 1$, and in particular $= 1$ in $(1) \ast\!\!\!\times \prod\prod_{\substack{\alpha \in I \\ \alpha \neq \alpha_0}} \Sigma_2$ and in $= -1$ in $0 \ast\!\!\!\times \prod\prod_{\substack{\alpha \in I \\ \alpha \neq \alpha_0}} \Sigma_2$.

This proves that $F_{\alpha_0}(x)$ is measurable, and that for any finite subset of I $\prod_{\alpha_0 \in \phi} F_{\alpha_0}(x)$ is also measurable. Further $\left|\prod_{\alpha_0 \in \phi} F_{\alpha_0}(x)\right|^2 = 1$, so that $\int_D \left|\prod_{\alpha_0 \in \phi} F_{\alpha_0}(x)\right|^2 d\tilde{\mu}(M_x) = 1$; thus all $\prod_{\alpha_0 \in \phi} F_{\alpha_0}(x)$ belong to $\bar{S}_{D,\tilde{\mu}^*(M)}$ and are normalized. One can verify easily that $\int \prod_{\alpha_0 \in \phi} F_{\alpha_0}(x) d\tilde{\mu}(M_x) = 0$ if $\phi \neq 0$; but as $\prod_{\alpha_0 \in \phi} F_{\alpha_0}(x) \cdot \prod_{\alpha_0 \in \psi} F_{\alpha_0}(x) = \prod_{\alpha_0 \in (\phi+\psi)-\phi\psi} F_{\alpha_0}(x)$, this proves $\int_D \prod_{\alpha_0 \in \phi} F_{\alpha_0}(x) \cdot \overline{\prod_{\alpha_0 \in \psi} F_{\alpha_0}(x)} \cdot d\tilde{\mu}(M_x) = 0$ if $\phi \neq \psi$. Thus the $\prod_{\alpha_0 \in \phi} F_{\alpha_0}(x)$ form a normalized orthogonal system in $\bar{S}_{D,\tilde{\mu}^*(M)}$.

But this system is complete. This is shown if we prove that their linear aggregates are everywhere dense. Now we saw in the proof of Theorem 12'.6 that those of the $1_{\sum_{i=1}^{n} N_i}(x)$; $(n=0,1,2,\ldots;\ N_1,\ldots,N_n \in \mathcal{R})$ are everywhere dense; thus it suffices to show that each $1_{\sum_{i=1}^{n} N_i}(x)$ is a linear aggregate of certain of our functions.

Now each $N \in \mathcal{R}$ has obviously the form $M \divideontimes \prod\prod_{\alpha \in I-\phi} \Sigma_2$, where $M \in \prod\prod_{\alpha \in \phi_0} \Sigma_2$, and ϕ_0 is some finite subset of I. We can replace ϕ_0 here by any finite set $\psi \subset \phi_0 \subset I$. Thus each $\sum_{i=1}^{n} N_i$, $N_i \in \mathcal{R}$, has the same form. Because of the obvious additivity, it suffices to consider the case where M is a one-point set; let its element be $(x_\alpha;\ \alpha \in \phi_0)$, where $x_\alpha = 0, 1$. Then the set to be considered is $L = \prod\prod_{\alpha \in \phi} (x_\alpha) \divideontimes \prod\prod_{\alpha \in I-\phi_0} \Sigma_2$; or if we put $\phi_0 = \phi_+ + \phi_-$, where $x_\alpha = 0$ in ϕ_- and $x_\alpha = 1$ in ϕ_+, then we have:

$L = \prod\prod_{\alpha \in \phi_-} (0) \divideontimes \prod\prod_{\alpha \in \phi_+} (1) \divideontimes \prod\prod_{\alpha \in I-(\phi_+ + \phi_-)} \Sigma_2$. This formula implies

$$1_L(x) = \prod_{\alpha \in \phi_-} \tfrac{1}{2}(1 - \mathcal{F}_\alpha(x)) \cdot \prod_{\alpha \in \phi_+} \tfrac{1}{2}(1 + \mathcal{F}_\alpha(x))$$

which is clearly a linear aggregate of $\prod_{\alpha_0 \in \phi} F_{\alpha_0}(x)$'s. This completes the proof of completeness for the $\prod_{\alpha_0 \in \phi} \mathcal{F}_{\alpha_0}(x)$.

Thus the dimension of $\bar{S}_{D,\tilde{\mu}(M)}$ is the power of the set of all $\prod_{\alpha_0 \in \phi} F_{\alpha_0}(x)$, that is of the set of all finite subsets ϕ of I. This is 1 + power (I) + + (power (I))^2 + ... It is clear that this power is $\geq$ power (I); but as I is infinite, it is even = 1 + power (I) + power (I) + ... = 1 + $\aleph_0 \cdot$ power (I) = = 1 + power (I) = power (I). Hence the dimension $\bar{S}_{D,\tilde{\mu}(M)}$ is the arbitrarily prescribed (infinite) power (I).

Note that in spite of the non-separability of $\bar{S}_{D,\tilde{\mu}(M)}$ (if power (I) > $\aleph_0$), the total measure of D is finite:

$$\tilde{\mu}(D) = \tilde{\mu}(\prod_{\alpha \in I} \Sigma_2) = 1 .$$

Other important examples of functional spaces satisfying our postulates originate from the theory of almost periodic functions. Thus the H.Bohr almost-periodic functions fulfill A, B, C_2, D_2, E_2, while A. Besicovitch's generalization of these functions fulfills A, B, C_2, D_2, E_1. (The latter space is isomorphic with the complete extension of the former in the sense of Theorem 12.15. Its dimension in the power of the continuum, $\aleph$.) But we do not intend to give further details of this subject here. The reader may be referred to the monograph of A. Besicovitch, "Almost periodic functions", Cambridge, 1932. (Cf. in particular Chapter II there, and its appendix, pp.67-128; especially pp.109-112.)

CHAPTER XIII.

LINEAR OPERATORS.

It is assumed in this chapter that S satisfies Postulates A, B, and E_1.

Definition 13.1. *The direct product of* S *by itself,* $S \times S$, *is the set of all ordered pairs,* $\langle f, f' \rangle$, *where* f *and* f' *are elements of* S. *In* $S \times S$, $a\langle f, f' \rangle$ *is taken to be* $\langle af, af' \rangle$, $\langle f, f' \rangle + \langle g, g' \rangle$ *is taken to be* $\langle f + g, f' + g' \rangle$, *and* $(\langle f, f' \rangle, \langle g, g' \rangle)$ *is taken to be* $(f, g) + (f', g')$.

It is apparent that $S \times S$ satisfies Postulates A and B. If $\langle f_1, g_1 \rangle$, $\langle f_2, g_2 \rangle$, ... is a fundamental sequence of elements of $S \times S$, then it is readily seen that this sequence has the limit $\langle f, g \rangle$ in $S \times S$, where $f = \lim_{n \to \infty} f_n$ and $g = \lim_{n \to \infty} g_n$. Hence $S \times S$ satisfies Postulate E_1.

Definition 13.2. *An operator (in* S) *is a function* ϕ *which is defined over some subset* D *of* S *and which has one or more values* ϕf *in* S *corresponding to each element* f *of* D. *The set* D *is called the (definition) domain,* $D(\phi)$, *of* ϕ. *The set of all values* ϕf, $f \in D(\phi)$, *is called the range,* $R(\phi)$, *of* ϕ. *The set of all elements* $\langle f, \phi f \rangle$, $f \in D(\phi)$ *and* $\phi f \in R(\phi)$, *is called the graph,* $G(\phi)$, *of* ϕ. *Thus* $D(\phi) \subset S$, $R(\phi) \subset S$, *and* $G(\phi) \subset S \times S$.

To say that $\langle f, g \rangle \in G(\phi)$ thus means that $f \in D(\phi)$, $g \in R(\phi)$, ϕf exists, and one of its values is g.

Definition 13.3. *An operator* ϕ *is called linear if* $G(\phi)$ *is a l.m.*

THEOREM 13.1. *If* ϕ *is linear, then* $D(\phi)$ *and* $R(\phi)$ *are l.m.'s.*

Proof: If $\langle f, (\phi f)_0 \rangle$ and $\langle g, (\phi g)_0 \rangle$ are elements of $G(\phi)$, where $(\phi f)_0$ and $(\phi g)_0$ are particular values ϕf and ϕg, then $\langle f, (\phi f)_0 \rangle + \langle g, (\phi g)_0 \rangle = \langle f + g, (\phi f)_0 + (\phi g)_0 \rangle \in G(\phi)$, and $a\langle f, (\phi f)_0 \rangle = \langle af, a(\phi f)_0 \rangle \in G(\phi)$. Therefore $(f + g) \in D(\phi)$ and $af \in D(\phi)$, so that $D(\phi)$ is a l.m., $[(\phi f)_0 + (\phi g)_0] \in R(\phi)$ and $a(\phi f)_0 \in R(\phi)$, so that $R(\phi)$ is a l.m., $\phi(f + g)$ exists and one of its values is $(\phi f)_0 + (\phi g)_0$, and $\phi(af)$ exists and one of its values is $a(\phi f)_0$.

As every linear set contains the zero element, $\langle 0, 0 \rangle$ is in the graph of every linear operator.

Definition 13.4. *An operator ϕ is called single-valued (s.v.) if there is associated exactly one value ϕf with each element $f \in D(\phi)$.*

Thus, if ϕ is s.v. and if $\langle f, g_1 \rangle$ and $\langle f, g_2 \rangle$ are in $G(\phi)$, then $g_1 = g_2 = \phi f$.

THEOREM 13.2. *A necessary and sufficient condition that a linear operator ϕ be s.v. is that $\phi(0)$ have the unique value 0.*

Proof: By the remark above, 0 is one of the values $\phi(0)$. Hence the condition is necessary. To show that it is sufficient, suppose that $\langle f, g_1 \rangle$ and $\langle f, g_2 \rangle$ are elements of $G(\phi)$. Since ϕ is linear, $\langle f, g_1 \rangle - \langle f, g_2 \rangle = \langle 0, g_1 - g_2 \rangle \in G(\phi)$. Thus $g_1 - g_2$ is one of the values $\phi(0)$. Since the unique value of $\phi(0)$ is 0, $g_1 - g_2 = 0$, and $g_1 = g_2$. Therefore ϕ is s.v.

By an argument similar to that used in the preceding proof, it may be shown that if ϕ is linear and if ϕ has two different values associated with some element in $D(\phi)$, then ϕ has infinitely many values associated with each element in $D(\phi)$.

If ϕ is s.v. and linear, and if f and g are in $D(\phi)$, then $f + g$ and af are in $D(\phi)$, and $\phi(f + g) = \phi f + \phi g$ and $\phi(af) = a\phi f$.

Definition 13.5. *An operator $\emptyset$ is called continuous if it is continuous at every point of its domain* $D(\emptyset)$ (cf. Definition 12.6).

It is apparent that this definition is meaningful only when $\emptyset$ is s.v. Hence, if an operator is described as being continuous, it is also described as being s.v.

It is desirable at this point to make a digression from the general theory of operators in order to discuss in detail a particularly important special class of operators, the projections.

Definition 13.6. *If M is a c.l.m. in S, if* $f \in S$, *and if* $f = f_1 + f_2$, *where* $f_1 \in M$ *and* $f_2 \in \ominus M$ (by Theorem 12.23, the representation $f_1 + f_2$ is unique), *then* f_1 *is called the projection of f on M and the operation of projecting f on M is denoted by* $P_M f = f_1$. (Note that projections are defined only with respect to c.l.m.'s.)

THEOREM 13.3: *A necessary and sufficient condition that an operator E be a projection* P_M *is that* 1) *E is s.v., linear, with* $D(E) = S$, 2) $(Ef, g) = (f, Eg)$ *for every f and g in S, and* 3) $E^2 = E$, *where* E^2 *is defined to be EE. M is uniquely determined by E.*

Proof of necessity: Condition 1) is immediate. 2) If f and g are in S, then $f = f_1 + f_2$ and $g = g_1 + g_2$, where $f_1 \in M$, $f_2 \in \ominus M$, $g_1 \in M$ and $g_2 \in \ominus M$. Then $(Ef, g) = (f_1, g_1 + g_2) = (f_1, g_1)$ and $(f, Eg) = (f_1 + f_2, g_1) = (f_1, g_1)$. 3) $Ef = f_1$ and $E^2 f = E(Ef) = Ef_1 = f_1$ (the decomposition of $f_1 \in M$ being $f_1 = f_1 + 0$).

Proof of uniqueness of M: If there exists a c.l.m. M such that $E = P_M$, then, for any element f of S, $Ef = P_M f$. If $f \in M$, then $Ef = P_M f = f$; if f is not in M, then $Ef = P_M f \neq f$. Hence M is the set of all solutions of the equation $Ef = f$, and $R(E) \supset M$. Now let g be any element of S and let $Eg = h$. Then $E^2 g = Eh$. Since $E^2 = E$, $Eh = h$, and $h \in M$. Therefore $R(E) \subset M$, and thus

$R(E) = M$. This completes the proof.

Proof of sufficiency: Let M be the set of all solutions of the equation $Ef = f$. Then, as in the preceding proof, $M = R(E)$. Since E is linear, M is linear (Theorem 13.1). It will now be shown that E is continuous. By Schwarz's Lemma, $\|Ef\| \cdot \|f\| \geqq |(Ef, f)| = |(E^2f, f)| = |(Ef, Ef)| = (Ef, Ef) = \|Ef\|^2$ for every $f \in S$. If $\|Ef\| > 0$, this relation may be divided by $\|Ef\|$ with the result that $\|f\| \geqq \|Ef\|$. But this last condition obviously holds if $\|Ef\| = 0$, so that it holds for all $f \in S$. Hence, for any f and g in S, $\|E(f - g)\| = \|Ef - Eg\| \leqq \|f - g\|$, and E is continuous over S. Therefore M is closed, so that M is a c.l.m. It remains to show that $E = P_M$. Let f be any element of S. Then $f = Ef + (f - Ef)$. But Ef is in M. Since $E^2f = Ef$, $Ef - E^2f =$ If g is any element of S, $0 = (Ef - E^2f, g) = (f - Ef, Eg)$. If g runs through S, Eg runs through M. Hence $(f - Ef) \in \ominus M$ and $E = P_M$.

Remark 1: If E is an operator satisfying conditions 2) and 3) of the preceding theorem, then, since $(Ef, g) = (E^2f, g) = (Ef, Eg)$, it follows that $(Ef, Eg) = (Ef, g) = (f, Eg)$. Conversely, if E is such that $(Ef, Eg) = (Ef, g)$ for every f and g of S, then conditions 2) and 3) obtain, for $(Ef, g) = (Ef, Eg) = \overline{(Eg, Ef)} = \overline{(Eg, f)} = (f, Eg)$ and $(Ef, g) = (Ef, Eg) =$ (by the preceding equation) $(E(Ef), g) = (E^2f, g)$, so that $(Ef - E^2f, g) = 0$. If g is taken to be $Ef - E^2f$, then $Ef - E^2f = 0$ and $E^2f = Ef$ for all f. Hence $E^2 = E$.

Remark 2: It should be particularly noted that, from the proof of the preceding theorem, if $E = P_M$, then M is the set of all solutions of the equation $Ef = f$, and $M = R(E)$.

Remark 3: Since $(Ef, f) = (Ef, Ef) = \|Ef\|^2$, (Ef, f) is real and non-negative. But it was shown that $\|Ef\| \leqq \|f\|$. Hence $0 \leqq (Ef, f) \leqq \|f\|^2$.

Remark 4: Certain projections are of particular interest.

1) If $M = [0]$, $P_M = 0$.

2) If $M = S$, $P_M = 1$.

3) If $M = [\varphi]$, where $\|\varphi\| = 1$, $P_M f = P_{[\varphi]} f = (f, \varphi)\varphi$.

4) $P_{\ominus M} f = f - P_M f = (1 - P_M)f$, so that $P_{\ominus M} = 1 - P_M$.

The preceding remark has some noteworthy consequences.

Remark 5: If E is a projection, 1 - E is also a projection, and since $E = 1 - (1 - E)$, it follows that E is a projection when and only when 1 - E is a projection. (This may be verified by the following computation: $(1 - E)^2 = 1 - E - E + E^2 = 1 - E$.)

Remark 6: If $E = P_M$, then $1 - E = P_{\ominus M}$. Thus $\ominus M$ is the set of all solutions of the equation $(1 - E)f = f$, that is, of the equation $Ef = 0$. Moreover, $\ominus M = R(1 - E)$.

Remark 7: If $E = P_M$, then we saw that $Ef = 0$ as well as $\|Ef\| = 0$ are characteristic for $f \in \ominus M$. $Ef = f$ is characteristic for $f \in M$. Thus $\|Ef\| = \|f\|$ is necessary. But it is also sufficient, because it implies $\|(1 - E)f\|^2 = ((1 - E)f, f) = (f, f) - (Ef, f) = \|f\|^2 - \|Ef\|^2 = 0$, $(1 - E)f = 0$, $Ef = f$. Thus $Ef = f$ as well as $\|Ef\| = \|f\|$ is characteristic for $f \in M$.

THEOREM 13.4. If E and F are projections in S, then a necessary and sufficient condition 1) that EF be a projection is that EF = FE, 2) that E + F be a projection is that EF + FE = 0, another such condition being that EF = 0, and a third such condition being that FE = 0, 3) that E - F be a projection is that EF = F, another such condition being that FE = F.

Proof: It is obvious that EF, E + F, and E - F are linear, s.v., and defined over the whole of S.

Part 1) Since $(EFf, g) = (Ff, Eg) = (f, FEg)$, the condition $EF = FE$ is necessary and sufficient that $(EFf, g) = (f, EFg)$. Moreover, the condition

$EF = FE$ is sufficient that $(EF)^2 = EFEF = EEFF = EF$. Hence the condition $EF = FE$ is necessary and sufficient that EF be a projection.

Part 2) It follows immediately that $((E + F)f, g) = (f, (E + F)g)$. Since $(E + F)^2 = E^2 + EF + FE + F^2 = (E + F) + (EF + FE)$, the condition $EF + FE = 0$ is necessary and sufficient that $E + F$ be a projection. But from the condition $EF + FE = 0$ it follows that $E(EF + FE) - (EF + FE)E = E^2F - FE^2 = EF - FE = 0$, so that $EF = FE = 0$. Conversely, if $EF = 0$, then since 0 is a projection, EF is a projection and, by Part 1), $EF = FE$. Hence $EF + FE = 0$. Therefore the condition $EF = 0$ is necessary and sufficient that $E + F$ be a projection. Since $E + F$ is symmetric in E and F, the condition $FE = 0$ is also necessary and sufficient.

Part 3) $E - F$ is a projection when and only when $1 - (E - F) = (1 - E) + F$ is a projection. By the preceding part, either of the conditions $(1 - E)F = 0$ or $F(1 - E) = 0$ is necessary and sufficient that $E - F$ be a projection. This completes the proof of the theorem.

It follows from the above proof that EF is a projection if either $E + F$ or $E - F$ is a projection.

<u>THEOREM 13.5:</u> <u>If</u> $E = P_M$ <u>and</u> $F = P_N$, <u>then</u> $EF = P_{MN}$, $E + F = P_{[M,N]}$, <u>and</u> $E(1 - F) = P_{M(\ominus N)}$ <u>when</u> EF, $E + F$, <u>and</u> $E(1 - F)$ <u>are projections</u>.

Proof: It is obvious that if $EF = P_L$, then $L \subset MN$, for if f is any element of S, then $E(Ff) \in M$, $F(Ef) \in N$, and $EF = FE$. Conversely, if f is any element of MN, then $Ef = f$, $Ff = f$, and thus $EFf = Ef = f$, so that $MN \subset L$. Hence $L = MN$. Again, if $E + F = P_L$, then $L \subset \{M, N\}$, for if f is any element of S, then $(E + F)f = (Ef + Ff) \in \{M, N\}$. Conversely, if f is any element of $\{M, N\}$, then $f = g + h$, where $g \in M$ and $h \in N$. Hence $(E + F)(g + h) = Eg + Fg + Eh + Fh$. But $Eg = g$ and $Fh = h$, so that $(E + F)(g + h) = g + h + FEg + EFh = g + h$ since $EF = FE = 0$. Hence $L \supset \{M, N\}$, and therefore $L = \{M, N\}$. But L is closed, so that $\{M, N\} = [M, N]$. The rest of the theorem is immediate.

THEOREM 13.6. If $E = P_M$ and $F = P_N$ and $EF = FE$, then $\{M, N\} = [M, N]$ and $E + F - EF = P_{[M,N]}$.

Proof: In the proof of Theorem 13.5 it was shown that $\{M, N\} = [M, N]$ if $EF = FE = 0$. But now the hypothesis is merely that $EF = FE$. Since $E(1 - F) = (1 - F)E$, therefore $E' = E(1 - F)$ is a projection, say $P_{M'}$. Since $E'F = 0$, $E' + F$ is a projection and, by Theorem 13.5, $\{M', N\} = [M', N]$. But EF and $E' + EF = E$ are projections, and $\{M', MN\} = [M', MN] = M$. With the aid of these two representations of M, it follows that, since $MN \subset N$, $\{M, N\} = \{M', MN, N\} = \{M', N\}$ and $[M, N] = [M', MN, N] = [M', N]$. But $\{M', N\} = [M', N]$. Hence $\{M, N\} = [M, N]$.

By Theorems 12.22 and 12.24 it follows that $[M, N] = \ominus(\ominus M \ominus N)$. Hence $P_{[M,N]} = 1 - (1 - P_M)(1 - P_N) = 1 - (1 - E)(1 - F) = E + F - EF$.

In order to generalize the preceding theorem to the case where $EF \neq FE$ it is necessary to introduce the following defintion and prove the following theorem.

Definition 13.7: If $\phi_1, \phi_2, \ldots$ is a sequence Σ of s.v. operators, if f is an element of $\prod_{n=1}^{\infty} D(\phi_n)$ such that $\lim_{n\to\infty} \phi_n f$ exists, and if D is the set of all such elements f, then Σ is said to have a limit ϕ over D, and, for $f \in D = D(\phi)$, $\phi f = \lim_{n\to\infty} \phi_n f$.

THEOREM 13.7. If $E = P_M$ and $F = P_N$, then the sequence Σ_1 of operators E, FE, EFE, FEFE, ... has a limit G, the sequence Σ_2: F, EF, FEF, ... has the same limit G, and $G = P_{MN}$. (The condition $EF = FE$ need not hold.)

Proof: Let A_n be the nth operator of the sequence Σ_1. Then $(A_m f, A_n g) = (A_{m+n-\varepsilon} f, g)$, where $\varepsilon = 1$ if m and n have the same parity and $\varepsilon = 0$ if m and n have opposite parity. It must be shown that if f is any element of S, then $\lim_{n\to\infty} A_n f$ exists. But $\| A_m f - A_n f \|^2 = (A_m f - A_n f, A_m f - A_n f) =$

$= (A_mf, A_mf) - (A_mf, A_nf) - (A_nf, A_mf) + (A_nf, A_nf) = (A_{2m-1}f, f) + (A_{2n-1}f, f) - 2(A_{m+n-\varepsilon}f, f)$. (Since $m + n - \varepsilon$ is always odd, the last term of this expression may be written as $2(A_{2k-1}f, f)$.) If it can be shown that $\lim_{i\to\infty} (A_{2i-1}f, f)$ exists, then the limit (as $m\to\infty$ and $n\to\infty$) of this last expression exists and is zero, so that $\lim_{n\to\infty} A_nf$ exists. Now $(A_{2i-1}f, f) = (A_if, A_if) = \|A_if\|^2$. Likewise $\|A_{i+1}f\|^2 = (A_{2i+1}f, f)$. But $A_{i+1}f$ is either EA_if or FA_if, so that, by Remark 3, $\|A_{i+1}f\| \leqq \|A_if\|$. Hence $(A_{2i-1}f, f) \geqq (A_{2i+1}f, f)$. Therefore $(A_1f, f) \geqq (A_3f, f) \geqq (A_5f, f) \geqq \ldots \geqq 0$. Thus $\lim_{i\to\infty} (A_{2i-1}f, f)$ exists; therefore $\lim_{n\to\infty} A_nf$ exists; let it be denoted by f^*. If G is defined by the condition $Gf = f^*$, then $D(G) = S$ and G is s.v. It is obvious that G is linear. But $\lim_{m,n\to\infty} (A_mf, A_ng) = \lim_{m,n\to\infty} (A_{m+n-\varepsilon}f, g)$, so that $(Gf, Gg) = (Gf, g)$. By Remark 1 and Theorem 13.3, G is a projection P_L.

If f is an element of MN, then $Ef = Ff = f$, $A_nf = f$, and $Gf = f$. By Remark 2, $f \in L$, so that $MN \subset L$. Since $EA_{2i} = A_{2i+1}$ and $FA_{2i-1} = A_{2i}$, it follows that, if $i\to\infty$, $EG = G$ and $FG = G$ since E and F are continuous. Let g be any element of S and let $Gg = f$. Then $Ef = f$ and $Ff = f$, so that $R(G) = L \subset MN$. Thus $L = MN$.

By interchanging E and F in this argument, it is seen that $\sum_2$ has a limit $G' = P_{MN}$, so that $G = G'$.

<u>Corollary</u>: <u>If</u> $E = P_M$ <u>and</u> $F = P_N$, <u>then</u> $P_{[M,N]} = 1 - G'$, <u>where</u> G' <u>is the limit of the sequence</u> $(1 - E)$, $(1 - F)(1 - E)$, $(1 - E)(1 - F)(1 - E)$, ...

Proof: By Theorems 12.22 and 12.24, $[M,N] = \ominus(\ominus M \ominus N)$. Hence $P_{[M,N]} = P_{\ominus(\ominus M \ominus N)} = 1 - P_{\ominus M \ominus N} = 1 - G'$.

If $E = P_M$ and $F = P_N$ do not commute, it is still the case that $\{M, N\} = [M, N]$ in Euclidean spaces of finite dimension, but it is not necessarily the case in other spaces, such as Hilbert space.

If $E = P_M$ and $F = P_N$, then the condition $E \leqq F$ is taken to mean $M \subset N$. It is readily seen that this sign of inequality as applied to projections has all the usual properties of this symbol. $M \subset N$ means $Ff = f$ for all $f \in M$, that is for all $f = Eg$. That is : $FEg = Eg$ for all g, or: $FE = E$. Thus $E \leqq F$ is equivalent to $FE = E$, and to the equivalent statements in Theorem 13.4, 3).

The assertion that M and N commute is taken to mean that E and F commute, that is, that $EF = FE$. If M and N are orthogonal, $M \perp N$, then E and F are said to be orthogonal. Now $M \perp N$ means that $0 = (Ef, Fg) = (f, EFg)$ for every f and g in S. Since the relation holds for every f, it states that $EFg = 0$ for every g, that is, $EF = 0$. Thus the orthogonality of E and F is equivalent to $EF = 0$, and to the equivalent statements in Theorem 13.4, 2).

<u>THEOREM 13.8. The condition $E \leqq F$ is equivalent to the condition that $\|Ef\| \leqq \|Ff\|$ for every $f \in S$, where E and F are projections.</u>

Proof: If $E \leqq F$, then, by the above remark, $EF = E$. By Remark 3 following Theorem 13.3, $\|Ef\| = \|EFf\| \leqq \|Ff\|$. Conversely, the second condition implies the first, for let $E = P_M$ and let $F = P_N$. Then if $\|Ef\| \leqq \|Ff\|$ for every $f \in S$, then in the case where $f \in M$, $Ef = f$ and $\|Ff\| \geqq \|f\|$. Therefore, by Remark 3, $(Ff,f) = \|Ff\|^2 \geqq \|f\|^2 = (f, f)$ and $((1-F)f, f) \leqq 0$. Since F is a projection, $1 - F$ is a projection and $((1 - F)f) = \|(1 - F)f\|^2 \leqq 0$. Hence $(1 - F)f = 0$ and $Ff = f$, so that, by Remark 2, $f \in N$. Therefore $M \subset N$, that is, $E \leqq F$. This completes the proof of the theorem.

<u>Corollary: If $E \leqq F$, then $\|Ef\| = \|Ff\|$ is equivalent to $Ef = Ff$.</u>

Proof: As $F - E$, E, F are projections, we have $\|Ff - Ef\|^2 = \|(F - E)f\|^2 = ((F - E)f, f) = (Ff, f) - (Ef, f) = \|Ff\|^2 - \|Ef\|^2$, and thus $\|Ef\| = \|Ff\|$ means $Ff - Ef = 0$, $Ef = Ff$.

<u>THEOREM 13.9. If $E_1, \ldots, E_n$ are projections, then a necessary and suf-</u>

ficient condition that $\sum_{i=1}^{n} E_i$ be a projection is that $E_\rho E_\sigma = 0$ whenever $\rho \neq \sigma$.

Proof: It is apparent that $\sum_{i=1}^{n} E_i$ satisfies conditions 1) and 2) of Theorem 13.3. If $E_\rho E_\sigma = 0$ for $\rho \neq \sigma$, then $(\sum_{i=1}^{n} E_i)^2 = \sum_{i=1}^{n} (E_i^2) + \sum_{\substack{i,j=1 \\ i \neq j}}^{n} E_i E_j = \sum_{i=1}^{n} E_i$, so that condition 3) is also satisfied.

Thus the condition of the theorem is sufficient.

If $\sum_{i=1}^{n} E_i$ is a projection, then, for every $f \in S$, $\|f\|^2 \geqq \|(\sum_{i=1}^{n} E_i)f\|^2 = (\sum_{i=1}^{n} E_i f, f) = \sum_{i=1}^{n} \|E_i f\|^2 \geqq \|E_\rho f\|^2 + \|E_\sigma f\|^2$, where $\rho \neq \sigma$. Let g be any element of S and let $E_\sigma g = f$; then $E_\sigma f = f$. By the preceding inequality, $E_\rho f = 0$. Therefore $E_\rho E_\sigma g = 0$ for every g, so that $E_\rho E_\sigma = 0$ for $\rho \neq \sigma$. Thus the condition of the theorem is necessary.

THEOREM 13.10. If a sequence of projections $E_1, E_2, \ldots$ is such that either $E_1 \geqq E_2 \geqq \ldots$ or $E_1 \leqq E_2 \leqq \ldots$, then the sequence has a limit E which is a projection.

Proof: If $E_1 \geqq E_2 \geqq \ldots$, then by Theorem 13.8, $\|E_1 f\|^2 \geqq \|E_2 f\|^2 \geqq \ldots \geqq 0$, where f is any element of S. The sequence $\|E_1 f\|^2, \|E_2 f\|^2, \ldots$ therefore has a limit, so that there exists a member n_0 such that, for m and $n > n_0$, $|\|E_m f\|^2 - \|E_n f\|^2| < \varepsilon$. Since the E's are projections, this condition becomes $|(E_m f, f) - (E_n f, f)| < \varepsilon$ (cf. Remark 3), that is, $|((E_m - E_n)f, f)| < \varepsilon$ for m and $n > n_0$. If $m \leqq n$, $E_m \geqq E_n$, and, by the remark preceding Theorem 13.8, $E_m - E_n$ is a projection. If $m > n$, $E_n - E_m$ is a projection and $|((E_n - E_m)f, f)| < \varepsilon$. In any event, $\|(E_m - E_n)f\|^2 = \|E_m f - E_n f\|^2 < \varepsilon$ for m and $n > n_0$. Therefore $\lim_{i \to \infty} E_i f$ exists; let this limit be f^*. Let E be defined by the condition $Ef = f^*$. But E is s.v., linear, and defined over all S, and since $(E_i f, E_i g) = (E_i f, g)$ for all i, $(Ef, Eg) = (Ef, g)$ by continuity. Hence, by Remark 1 and Theorem 13.3, E is a projection, and the first part of the theorem is proved. The second part follows immediately from the first and the fact that if $E \leqq F$. $1 - E \geqq 1 - F$.

Corollary 1: In the preceding theorem, $E_1 \geqq E_2 \geqq \ldots \geqq E$ and $E_1 \leqq E_2 \leqq \ldots \leqq E$.

Corollary 2: In the preceding theorem, if F is a projection such that $E_i \geqq F$ or $E_i \leqq F$ for all i, then $E \geqq F$ or $E \leqq F$; hence E is the greatest (smallest) projection satisfying the conditions of the preceding corollary.

Corollary 3: In the preceding theorem, if $E_i = P_{M_i}$ and $E = P_M$, then $M = \prod_{i=1}^{\infty} M_i$ or $M = [M_1, M_2, \ldots]$.

Proofs: The proofs of all these corollaries are apparent.

It is now desirable to resume the development of the general theory of operators.

If $\varphi_1, \varphi_2, \ldots$ and $\varphi_1', \varphi_2', \ldots$ are two (not necessarily countable) complete o.n. sets in S, then the set of all elements $\langle \varphi_\alpha, 0 \rangle$ together with all elements $\langle 0, \varphi_\beta' \rangle$ is a complete o.n. set in $S \times S$. Thus the dimension of $S \times S$ (cf. Definition 12.7) is double the dimension of S. If it is infinite, therefore the two dimensions are equal (cf. Hausdorff, p.71).

Let X be the set of all elements $\langle f, 0 \rangle$ in $S \times S$ and let Y be the set of all elements $\langle 0, f \rangle$. Then X and Y are c.l.m.'s in $S \times S$. If $\langle g, g' \rangle \perp X$, then $(\langle f, 0 \rangle, \langle g, g' \rangle) = 0$, $(f, g) = 0$ for all $f \in S$, and $g = 0$. Hence $\ominus X = Y$, and, in the same way, $\ominus Y = X$. Let I_X be the isomorphism mapping S on X: $I_X f = \langle f, 0 \rangle$, and let I_Y be the isomorphism mapping S on Y: $I_Y f = \langle 0, f \rangle$. Then I_X^{-1} and I_Y^{-1} map X and Y on S. If $P_X \langle f, g \rangle$ is defined to be $\langle f, 0 \rangle$ for all g, and if $P_Y \langle f, g \rangle$ is defined to be $\langle 0, g \rangle$ for all f, then P_X and P_Y are projections of $S \times S$ on X and Y. It follows from this discussion that $D(\phi) = I_X^{-1}(P_X G(\phi))$ and $R(\phi) = I_Y^{-1}(P_Y G(\phi))$; conversely, $G(\phi)$ is the set of all sums $\langle f, 0 \rangle + \langle 0, g \rangle$, where $f \in D(\phi)$ and g is one of the values ϕf.

Definition 13.8: An operator ϕ is called closed if $G(\phi)$ is closed.

Let $f_1, f_2, \ldots$ be a sequence of elements of S and consider the following five conditions: 1) $\lim_{n \to \infty} f_n = f$, 2) $f_n \in D(\phi)$ for each n, 3) $\lim_{n \to \infty} (\phi f_n)_o$ exists (denote it by f_o), where $(\phi f_n)_o$ is one of the values ϕf_n, 4) $f \in D(\phi)$, and 5) f_o is one of the values ϕf.

THEOREM 13.11. That ϕ is closed means that conditions 1), 2), and 3) (in the preceding paragraph) together imply 4) and 5); that ϕ is continuous means that 1), 2), and 4) together imply 3) and 5) (where now 5) means that $\phi f = f_o$); if ϕ is linear, then the continuity of ϕ is sufficient that 1) and 2) together imply 3). Hence if ϕ is closed, linear, and continuous, its domain is closed.

Proof: To say that $G(\phi)$ is closed means that, if $P_1, P_2, \ldots$ is any fundamental sequence in $G(\phi)$, then the sequence has a limit in $G(\phi)$. But the statement that $P_1, P_2, \ldots$ is a fundamental sequence in $G(\phi)$ is equivalent to conditions 1), 2), and 3), and the statement that the limit is in $G(\phi)$ is equivalent to conditions 4) and 5). The second statement follows immediately from Definition 13.5. The third statement results by applying the second to the double sequence $f_m - f_n$, where it is remembered that S is complete.

It should be noticed that the closedness of ϕ does not imply the closedness of either $D(\phi)$ or $R(\phi)$.

Definition 13.9: If $G(\phi) \subset G(P)$, then P is called a continuation of ϕ and ϕ is called a contraction of P; this relation between ϕ and P is indicated by the notation $\phi \subset P$.

Definition 13.10: $\hat{\phi}$ is that operator whose graph $G(\hat{\phi})$ is $\{G(\phi)\}$; $\tilde{\phi}$ is that operator whose graph $G(\tilde{\phi})$ is $[G(\phi)]$. Thus $\tilde{\phi} \supset \hat{\phi} \supset \phi$.

This definition may be reformulated as follows: $D(\hat{\phi})$ is the set of all elements $a_1 f_1 + \ldots + a_n f_n$, where $f_1, \ldots, f_n \in D(\phi)$, and $\hat{\phi}(a_1 f_1 + \ldots a_n f_n)$ is the set of all values $a_1\emptyset f_1 + \dots + a_n\emptyset f_n$. $D(\tilde{\emptyset})$ is the set of all elements f such that 1) there exists a sequence $f_1, f_2, \dots$ of elements in $D(\hat{\emptyset})$ with limit f, and 2) it is possible to select a set of values $(\hat{\emptyset}f_1)_o, (\hat{\emptyset}f_2)_o, \dots$ such that $\lim_{n\to\infty} (\hat{\emptyset}f_n)_o$ exists. (By $(\hat{\emptyset}f_n)_o$ is meant one of the values $\hat{\emptyset}f_n$.) Let $f^\circ = \lim_{n\to\infty} (\hat{\emptyset}f_n)_o$. $\tilde{\emptyset}f$ is the set of all possible values f° arising from all possible sequences $f_1, f_2, \dots$ of the sort described.

For any operator $\emptyset$, $\hat{\emptyset}$ and $\tilde{\emptyset}$ are linear. (Cf. Theorem 13.1). If an operator is s.v., then each of its contractions is s.v.; for example, if $\tilde{\emptyset}$ is s.v., then $\emptyset$ is s.v., and if $\tilde{\emptyset}$ is s.v., then $\hat{\emptyset}$ and $\emptyset$ are s.v. But if $\emptyset$ is s.v., then in general neither $\hat{\emptyset}$ nor $\tilde{\emptyset}$ is s.v., and if $\hat{\emptyset}$ is s.v., then in general $\tilde{\emptyset}$ is not s.v. As mentioned above, $D(\tilde{\emptyset})$ may not be closed.

Definition 13.11. The isomorphism $\bar{U}\langle f, f'\rangle = \langle -f', f\rangle$ maps $S \times S$ on itself and will always be denoted by $\bar{U}$.

It is obvious that $\bar{U}$ is linear, that $\bar{U}^2 = -1$, and that $\bar{U}^4 = 1$. Thus $\bar{U}^2M = -M$, and if M is linear, $\bar{U}^2M = M$. If F and G are in $S \times S$, Then $(\bar{U}F, \bar{U}G) = (F, G)$. Thus $\|\bar{U}F\| = \|F\|$. The condition $M \perp N$ implies that $\bar{U}M \perp \bar{U}N$. Finally, $\ominus(\bar{U}M) = \bar{U}(\ominus M)$.

Definition 13.12. If $\bar{U}G(\emptyset) \perp G(P)$, then $\emptyset$ is said to be partially adjoint to P.

THEOREM 13.12. _A necessary and sufficient condition that $\emptyset$ be partially adjoint to P is that $(\emptyset f, g) = (f, Pg)$ for every $f \in D(\emptyset)$, every $g \in D(P)$, and all values $\emptyset f$ and Pg._

Proof: If $f \in D(\emptyset)$ and $g \in D(P)$, then, for any values $\emptyset f$ and Pg, $\bar{U}\langle f, \emptyset f\rangle \perp \langle g, Pg\rangle$, $\langle -\emptyset f, f\rangle \perp \langle g, Pg\rangle$, $(\langle -\emptyset f, f\rangle, \langle g, Pg\rangle) = 0$, $-(\emptyset f, g) + (f, Pg) = 0$, and the condition is necessary. A reversal of the argument shows that the condition is sufficient.

Corollary: If $\emptyset$ is partially adjoint to P, then P is partially adjoint to ϕ.

Proof: By Theorem 13.12, $(\phi f, g) = (f, Pg)$. Hence $(\overline{\phi f, g}) = (\overline{f, Pg})$, that is, $(Pg, f) = (g, \phi f)$.

If $\bar{U}G(\phi) \perp G(P)$, then $G(P) \subset \ominus (\bar{U}G(\phi))$. Hence if P is such that $G(P) = \ominus \bar{U}G(\phi)$, then P is the maximal operator partially adjoint to ϕ. This leads to

Definition 13.13. The operator ϕ^* such that $G(\phi^*) = \ominus \bar{U}G(\phi)$ is called the adjoint of ϕ.

By Theorem 12.22, every adjoint is closed and linear.

Since $\bar{U}G(\phi) \perp G(\phi^*)$, it follows as in the proof of Theorem 13.12 that $(\phi f, g) = (f, \phi^* g)$ whenever ϕf and $\phi^* g$ have sense. Conversely, $D(\phi^*)$ is the set of all elements g of S for which there exists an element g^* of S such that $(\phi f, g) = (g, g^*)$ for every $f \in D(\phi)$ and all values ϕf, and $R(\phi^*)$ is the set of all g^* corresponding to g.

If $\phi \subset P$, then, by Theorem 1.22, $\phi^* \supset P^*$. Since $G(\hat{\phi}) = \{G(\phi)\}$ and $G(\tilde{\phi}) = [G(\phi)]$, the same theorem shows that $\phi^* = \hat{\phi}^* = \tilde{\phi}^*$.

THEOREM 13.13. $\phi^{**} = \tilde{\phi} \supset \phi$.

Proof: By Theorem 12.24, $G(\phi^{**}) = \ominus \bar{U}(\ominus \bar{U}G(\phi)) = \ominus \bar{U}(\bar{U} \ominus G(\phi)) = \ominus \bar{U}^2 \ominus G(\phi) = \ominus (-(\ominus G(\phi))) = \ominus \ominus G(\phi) = [G(\phi)] = G(\tilde{\phi})$.

THEOREM 13.14: ϕ^* is s.v. when and only when $[D(\phi)] = S$.

Proof: Since ϕ^* is linear, it follows by Theorem 13.2 that it is sufficient to show that $\phi^*(0)$ has the unique value 0 when and only when $[D(\phi)] = S$. Suppose that h is any value of $\phi^*(0)$. This means that $\langle 0, h\rangle \perp \bar{U}G(\phi)$, $\bar{U}\langle 0, h\rangle \perp \bar{U}^2 G(\phi)$, and $\langle -h, 0\rangle \perp -G(\phi)$. Thus for any $f \in D(\phi)$ and any value ϕf, $(\langle -h, 0\rangle, \langle -f, -\phi f\rangle) = 0$ and $(h, f) = 0$. Hence $h \perp D(\phi)$, $h \perp [D(\phi)]$, and $h \in \ominus [D(\phi)]$. Therefore $h = 0$ is the unique solution of these conditions (i.e., ϕ^* is s.v.) when and only when $\ominus [D(\phi)] = (0)$. But this condition is equivalent to the condition that $\ominus\ominus [D(\phi)] = \ominus (0)$, that is, that $[D(\phi)] = S$.

(Cf. Theorem 12.24.)

Corollary 1. If ϕ is linear, then ϕ^* is s.v. when and only when $D(\phi)$ *is dense in* S.

Corollary 2: $\tilde{\phi}$ is s.v. when and only when $D(\phi^*)$ *is dense in* S.

Corollary 3: If P is partially adjoint to ϕ and if $[D(P)] = S$, *then $\tilde{\phi}$ is s.v.*

Corollary 4: In Corollary 3, if $D(\phi) = S$, *then* $\phi = P^*$ *and* $\tilde{P} = \phi^*$.

Corollary 5: $\phi \in \Psi$ *implies* $\phi^* \supset \Psi^*$.

Proofs: Corollary 1 follows from the fact that $D(\phi)$ is linear. Corollary 2 follows from Theorems 13.13 and 13.14 and the linearity of ϕ^*. Corollary 3 follows from Corollary 2 and the fact that $\phi^* \supset P$. With regard to Corollary 4: by Theorem 13.14, ϕ^* and P^* are s.v. Since $\phi^* \supset P$, $\phi^{**} \subset P^*$ and $\phi \subset P^*$. Since $D(\phi) = S$, $\phi = P^*$. Hence $\phi^* = P^{**} = \tilde{P}$. Corollary 5 is obvious.

Definition 13.14: ϕ is called Hermitian if ϕ is partially adjoint to itself, that is, if $(\phi f, g) = (f, \phi g)$ *for every* f *and* g *in* $D(\phi)$; *ϕ is called self-adjoint (s.a.) if* $\phi = \phi^*$.

THEOREM 13.15: If ϕ is s.a., then ϕ is Hermitian.

Proof: $(\phi f, g) = (f, \phi^* g) = (f, \phi g)$.

THEOREM 13.16: If ϕ is Hermitian, then $\phi^* \supset \phi^{**} \supset \phi$.

Proof: $\phi^* \supset \phi$ since ϕ^* contains any partial adjoint of ϕ. Hence $\phi^{**} \subset \phi^*$; since $\phi^{**} \supset \phi$, $\phi^* \supset \phi^{**} \supset \phi$.

Corollary: If ϕ is s.a., then $\phi^* = \phi^{**} = \phi$.

Proof: The proof is apparent.

THEOREM 13.17. The condition $\phi^* = \phi^{**}$ *is equivalent to the condition that $\tilde{\phi}$ is s.a.; the condition* $\phi = \phi^{**}$ *implies that ϕ is closed and linear.*

Proof: From the condition $\phi^* = \phi^{**}$ it follows that $\phi^{**} = \phi^{***}$, i.e., that $\tilde{\emptyset} = \tilde{\emptyset}^*$. Conversely, if $\tilde{\emptyset} = \tilde{\emptyset}^*$, then $\tilde{\emptyset}^* = \tilde{\emptyset}^{**}$. Since $\tilde{\emptyset}^* = \emptyset^*$, it follows that $\emptyset^* = \emptyset^{**}$. The rest of the theorem follows from Theorem 13.13.

THEOREM 13.18: If $\emptyset$ is Hermitian and $D(\emptyset) = S$, then $\emptyset$ is s.a. and s.v.

Proof: This is the special case of Corollary 4 of Theorem 13.14 with $P=\emptyset$.

It follows from this theorem that every projection is s.a.

Definition 13.15: $\emptyset$ is called bounded over $M \subset D(\emptyset)$ if there exists a number C such that $\|\emptyset f\| \leq C$ for all $f \in M$ and for all values $\emptyset f$.

THEOREM 13.19: If $\emptyset$ is linear, and if there exists an open sphere $\mathcal{T}$ in S such that $\emptyset$ is bounded over $\mathcal{T} \cdot D(\emptyset)$ and $\mathcal{T} \cdot D(\emptyset)$ is not empty, then $\emptyset$ is continuous (and therefore s.v.) over $D(\emptyset)$. (Cf. Definition 13.5 and remark following.)

Proof: Suppose $f_0 \in \mathcal{T} \cdot D(\emptyset)$. ($f_0$ exists.) Then there exists a closed sphere $\mathcal{T}' \subset \mathcal{T}$ with center f_0. Let the radius of $\mathcal{T}'$ be ε. If g is any element of $D(\emptyset)$ such that $\|g\| \leq \varepsilon$, then $f = f_0 \pm g$ is in $\mathcal{T}' \cdot D(\emptyset)$ and $\|\emptyset(f_0 \pm g)\| \leq C$, where C is a bound of $\emptyset$ over $\mathcal{T} \cdot D(\emptyset)$. Since $\emptyset$ is linear, $\|\emptyset(2g)\| = \|\emptyset(f_0 + g) - \emptyset(f_0 - g)\| \leq \|\emptyset(f_0 + g)\| + \|\emptyset(f_0 - g)\| \leq 2C$. Hence $\|\emptyset g\| \leq C$. If $\emptyset(0)$ has a value $(\emptyset 0)_0$, every $\alpha(\emptyset 0)_0$ is also a value of $\emptyset(0)$ by linearity. Thus $\|\alpha(\emptyset 0)_0\| \leq C$, $\|(\emptyset 0)_0\| \leq \frac{C}{|\alpha|}$ for every complex α. This requires that $(\emptyset 0)_0 = 0$. Since $\emptyset$ is linear, $\emptyset$ is s.v.

Let h be any element of $D(\emptyset)$ and if $h \neq 0$ let $h_1 = \frac{\varepsilon}{\|h\|} h$. Then $\|h_1\| = \varepsilon$, $\|\emptyset h_1\| = \frac{\varepsilon}{\|h\|} \|\emptyset h\| \leq C$, and $\|\emptyset h\| \leq \frac{C}{\varepsilon} \|h\|$. For $h = 0$ the last inequality still holds. If $h = p - p_0$, where $p \in D(\emptyset)$ and $p_0 \in D(\emptyset)$, then $\|\emptyset p - \emptyset p_0\| \leq \frac{C}{\varepsilon} \|p - p_0\|$ and $\emptyset$ is continuous over $D(\emptyset)$.

Let $\emptyset$ be a linear operator, and consider the following six conditions:

a) There exists an open sphere $\mathcal{T}$ in S such that $\emptyset$ is bounded over $\mathcal{T} \cdot D(\emptyset)$ and $\mathcal{T} \cdot D(\emptyset)$ is not empty.

b) There exists a constant $A \geqq 0$ such that $\|\phi f\| \leqq A\|f\|$, $f \in D(\phi)$.

c) There exists a constant $A \geqq 0$ such that $\|\phi f - \phi g\| \leqq A\|f - g\|$, f and g in $D(\phi)$.

d) ϕ is continuous and s.v. over $D(\phi)$.

e) $|(\phi f, g)| \leqq A\|f\| \cdot \|g\|$, $f \in D(\phi)$ and $g \in S$.

e') $|(\phi f, f)| \leqq A\|f\|^2$, $f \in D(\phi)$.

THEOREM 13.20: If ϕ is a linear operator, then the five conditions a) - e) are equivalent to each other. If ϕ is linear, Hermitian, with $D(\phi)$ dense in S, a) - e') are equivalent.

Proof: The proof of Theorem 13.19 shows that a) implies b), c), and d); it is obvious that each of these latter conditions implies a). Hence the first four of these conditions are equivalent. It is apparent that e) implies e'), and that b) implies e) (by Schwarz's Lemma). To show that e) implies b), let $g = \phi f$. Then $(\phi f, \phi f) = \|\phi f\|^2 \leqq A\|f\| \cdot \|\phi f\|$, and b) is immediate. To show that e') implies e) when ϕ is Hermitian and $D(\phi)$ is dense in S, replace f by $\frac{f \pm g}{2}$, where f and g are in $D(\phi)$, and take the difference between the two results. Then $\frac{1}{4}[(\phi f + \phi g, f + g) - (\phi f - \phi g, f - g)] = \mathcal{R}(\phi f, g) \leqq$ $\leqq \frac{1}{4}A[2\|f\|^2 + 2\|g\|^2]$. If, as in the proof of Schwarz's Lemma, f and g are replaced by af and $\frac{1}{a}$ g, and then f replaced by θf, $|\theta| = 1$, it follows, exactly as in the proof mentioned, that $|(\phi f, g)| \leqq A\|f\| \cdot \|g\|$. So far, f and g must both belong to $D(\phi)$. Since $D(\phi)$ is dense in S, it follows by continuity that this relation holds for any $g \in S$. Thus conditions a) to e) are equivalent, and if ϕ is Hermitian with $D(\phi)$ dense in S, then conditions a) to e') are equivalent.

Definition 13.16: If a linear operator satisfies the above five (six) conditions it is called bounded.

Thus, if ϕ is bounded, it is linear, continuous, and s.v. over $D(\phi)$.

It is apparent from the above proof that b), c), e) and e') hold for the same set of A's. Let A_0 be the g.l.b. of these A's. Then b), c), e) and e') hold for all $A \geqq A_0$. It is apparent that A_0 is the l.u.b. of each of the following expressions: $\frac{\|\phi f\|}{\|f\|}$, $\frac{\|\phi f - \phi g\|}{\|f - g\|}$, $\frac{|(\phi f, g)|}{\|f\| \cdot \|g\|}$, $\left(\frac{|(\phi f, f)|}{\|f\|^2}\right)$.

DEFINITION 13.17: $|||\phi|||$ is taken to be A_0; if ϕ is linear, Hermitian, and $D(\phi)$ is dense in S, then $\underline{|||\phi|||} = \text{g.l.b.} \frac{(\phi f, f)}{\|f\|^2}$ and $\overline{|||\phi|||} = \text{l.u.b.} \frac{(\phi f, f)}{\|f\|^2}$ ($(\phi f, f)$ is real); if, furthermore, $(\phi f, f) \geqq 0$ for all f in $D(\phi)$, ϕ is called semi-definite, and if $(\phi f, f) = 0$ only when $f = 0$, then ϕ is called definite.

It follows that, if ϕ is Hermitian, $|||\phi||| = \max\{|\underline{|||\phi|||}|, |\overline{|||\phi|||}|\}$.

The next three theorems are concerned with the question as to how much can be said about $\tilde{\phi}$ and ϕ^* from the properties of ϕ.

THEOREM 13.21: If ϕ is bounded with $D(\phi)$ dense in S, then $|||\phi||| = |||\tilde{\phi}||| = |||\phi^*|||$, $D(\tilde{\phi}) = D(\phi^*) = S$, and $\tilde{\phi}$ and ϕ^* are continuous and s.v. over the whole of S.

Proof: Since $\tilde{\phi} \supset \phi$, $(A_0)_{\tilde{\phi}} \geqq (A_0)_{\phi}$. Now let $f_1, f_2, \ldots$ be any fundamental sequence in $D(\phi)$ such that $\lim_{n\to\infty} \phi f_n$ exists. Let f and $\tilde{\phi} f$ be the limits of these sequences. If, for all n, $A \geqq \frac{\|\phi f_n\|}{\|f_n\|}$, then $A \geqq \frac{\|\tilde{\phi} f\|}{\|f\|}$. Hence $(A_0)_{\phi} \geqq (A_0)_{\tilde{\phi}}$, and $|||\phi||| = |||\tilde{\phi}|||$. Thus $\tilde{\phi}$ is bounded, and hence continuous and s.v. over $D(\phi)$ which is dense in S. By Corollary 2 of Theorem 13.14, $D(\phi^*)$ is dense in S. Hence, by continuity, the condition $|(\phi f, g)| \leqq A\|f\| \cdot \|g\|$ for $f \in D(\phi)$ and $g \in S$ is not weakened if g is restricted to being in $D(\phi^*)$. But $|(\phi f, g)| = |(g, \phi f)| = |(\phi^* g, f)|$. Hence $|(\phi^* g, f)| \leqq A\|f\| \cdot \|g\|$ for $f \in D(\phi)$ and $g \in D(\phi^*)$. Since $D(\phi)$ is dense in S, this condition still holds, by continuity, for $f \in S$. Therefore $|||\phi||| = |||\phi^*|||$, so that ϕ^* is bounded, continuous and s.v. over $D(\phi^*)$ which is dense in S. Since $\tilde{\phi}$ and ϕ^* are linear, closed, and continuous, it follows by Theorem 13.11 that $D(\tilde{\phi}) = D(\phi^*) = S$.

Lemma: If ϕ is linear, closed, s.v., discontinuous, and $D(\phi) = S$, if C is any positive number, and if γ is any closed sphere in S, then there exists a closed sphere γ_1 in γ over which $\|\phi f\| > C$.

Proof: Since ϕ is discontinuous, there exists an interior element $f_o \in \gamma$, such that $\|\phi f_o\| > C$. Then we have for $g_o = \phi f_o \neq 0$, $\frac{(\phi f_o, g_o)}{\|g_o\|} = \frac{(\phi f_o, \phi f_o)}{\|\phi f_o\|} = \|\phi f_o\| > C$. By Corollary 2 of Theorem 13.14, $D(\phi^*)$ is dense in S. Since $\frac{(\phi f_o, g)}{\|g\|}$ is continuous in g, this implies the existence of a $g_1 \in D(\phi^*)$ with $\frac{(\phi f_o, g_1)}{\|g_1\|} > C$. Now $\frac{(\phi f, g_1)}{\|g_1\|} = \frac{(f, \phi^* g_1)}{\|g_1\|}$ is continuous in f; therefore there exists an $\eta > 0$, such that $\|f - f_o\| \leqq \eta$ implies $\frac{(\phi f, g_1)}{\|g_1\|} > C$. Considering $\frac{(\phi f, g_1)}{\|g_1\|} \leqq \frac{\|\phi f\| \cdot \|g_1\|}{\|g_1\|} = \|\phi f\|$, we even have $\|\phi f\| > C$. Thus the closed sphere γ_1: $\|f - f_o\| \leqq \eta$ meets our requirements, as η can be made sufficiently small that $\gamma_1 \subset \gamma$.

THEOREM 13.22. If ϕ is linear, closed, s.v. and $D(\phi) = S$, then ϕ is continuous over the whole of S.

Proof: Suppose ϕ were discontinuous. By the preceding lemma it follows that, if γ_o is any closed sphere in S, there exists a closed sphere $\gamma_1 \subset \gamma_0$ such that $\|\phi f\| > C_1 = 1$ over γ_1, and its radius may be taken $< \frac{1}{1}$. Again there exists a closed sphere $\gamma_2 \subset \gamma_1$ such that $\|\phi f\| > C_2 = 2$ over γ_2, and its radius may be taken $< \frac{1}{2}$. Repetition of this argument shows that there exists a sequence of closed spheres $\gamma_0 \supset \gamma_1 \supset \gamma_2 \supset \ldots$ such that $\|\phi f\| > n$ over γ_n, and the radius of γ_n is $\frac{1}{n}$. As S is complete, there exists at least one element f_o in all the spheres γ_n and $\|\phi f_o\|$ is not finite. This contradicts the hypothesis that ϕ is s.v. over S. Hence ϕ is continuous.

THEOREM 13.23. If ϕ is linear, closed, s.v., with $D(\phi)$ dense in S, then ϕ^* has the same properties.

Proof: The first two of these properties are apparent. The third follows from Corollary 1 of Theorem 13.14. The fourth follows from Corollary 2 of Theorem 13.14 since $\phi = \tilde{\phi}$.

THEOREM 13.24. If ϕ is Hermitian, if $[D(\phi)] = S$, and if $\phi^2 \supset \phi$, then $\tilde{\phi}$ is a projection.

Proof: By Corollary 3 of Theorem 13.14, $\tilde{\phi}$ is s.v.; hence ϕ and ϕ^2 are s.v. Since $\phi^2 \supset \phi$, $\phi^2 = \phi$ over $D(\phi)$. Since ϕ is Hermitian, $(\phi f, \phi g) = (\phi^2 f, g) = (\phi f, g)$ for all f and g in $D(\phi)$. Hence $(\hat{\phi} f, \hat{\phi} g) = (\hat{\phi} f, g)$ for all f and g in $D(\hat{\phi})$, and, by continuity, $(\tilde{\phi} f, \tilde{\phi} g) = (\tilde{\phi} f, g)$ for all f and g in $D(\tilde{\phi})$ (where it is recalled that $\tilde{\phi}$ (and $\hat{\phi}$) is s.v.). Therefore $\|\tilde{\phi} f\|^2 = (\tilde{\phi} f, f) \leqq \|\tilde{\phi} f\| \cdot \|f\|$, and $\|\phi f\| \leqq \|f\|$ for all f in $D(\tilde{\phi})$. Therefore $\|\tilde{\phi}(g - h)\| = \|\tilde{\phi} g - \tilde{\phi} h\| \leqq \|g - h\|$ for all g and h in $D(\tilde{\phi})$, and $\tilde{\phi}$ is continuous. Since $\tilde{\phi}$ is s.v., it is bounded; $\tilde{\phi}$ is linear; since $[D(\tilde{\phi})] = S$, it follows by Theorem 13.21 that $D(\tilde{\phi}) = S$. The theorem follows from Remark 1 and Theorem 13.3.

Definition 13.18: If ϕ is an operator, if f is a given element of S, if there exists a $g \varepsilon D(\phi)$ such that one of the values of ϕg is f, then the set of values of $\phi^{-1} f$ is taken to be the set of all such g's.

It follows immediately that $D(\phi^{-1}) = R(\phi)$, $R(\phi^{-1}) = D(\phi)$, and $(\phi^{-1})^{-1} = \phi$. If $\phi \subset P$, then $\phi^{-1} \subset P^{-1}$.

THEOREM 13.25: ϕ and ϕ^{-1} have the same character with regard to linearity, closedness, Hermiticity, and self-adjointness; if ϕ and P are (partially) adjoint, then ϕ^{-1} and P^{-1} are also.

Proof: It is apparent that $G(\phi^{-1}) = \bar{U} G(-\phi)$. Hence $G(\widehat{\phi^{-1}}) = \{\bar{U} G(-\phi)\} = \bar{U}\{G(-\phi)\} = \bar{U} G(\widehat{-\phi}) = \bar{U} G(-\hat{\phi}) = G((\hat{\phi})^{-1})$. Similarly, $\widetilde{\phi^{-1}} = (\tilde{\phi})^{-1}$ and $(\phi^{-1})^* = (\phi^*)$. These three relations prove the theorem with regard to linearity, closedness, and self-adjointness. If $(\phi f, g) = (f, Pg)$, then $(\phi^{-1} f, g) = (\phi^{-1} f, PP^{-1} g) = (\phi\phi^{-1} f, P^{-1} g) = (f, P^{-1} g)$. Thus if ϕ and P are partially adjoint, ϕ^{-1} and P^{-1} are also. The remaining parts of the theorem result in a similar way.

The remaining theorems of this chapter have special applications to certain operators to be discussed later.

THEOREM 13.26. *If ϕ is linear, closed, s.v., with $D(\phi)$ dense in S, then $\phi^*\phi$ and $\phi\phi^*$ are s.a., linear, closed, s.v., with domains dense in S.*

Proof: By Theorem 13.23, ϕ^* is s.v. Since $G(\phi)$ is a c.l.m., any element F in $S \times S$ can be represented as $F = F_1 + F_2$, where $F_1 \in G(\phi)$ and $F_2 \in \ominus G(\phi)$. Since $G(\phi^*) = \ominus \bar{U}G(\phi)$, $\bar{U}G(\phi^*) = \ominus(-G(\phi)) = \ominus G(\phi)$. Hence $F_2 \in \bar{U}G(\phi^*)$. If F is the element $\langle f, 0 \rangle$, then F_1 is an element $\langle g, \phi g \rangle$, $g \in D(\phi)$, and F_2 is an element $\langle -\phi^* h, h \rangle$, $h \in D(\phi^*)$. Thus $f = g - \phi^* h$ and $0 = \phi g + h$, so that $h = -\phi g$. Hence $f = (\phi^*\phi + 1)g$. Since f is an arbitrary element in S, $R(\phi^*\phi + 1) = D((\phi^*\phi + 1)^{-1}) = S$. If ϕ is replaced by ϕ^*, then, since $\phi^{**} = \phi$, $R(\phi\phi^* + 1) = D((\phi\phi^* + 1)^{-1}) = S$. But $((\phi^*\phi + 1)f, g) = (\phi^*\phi f, g) + (f, g) = (\phi f, \phi g) + (f, g) = (f, \phi^*\phi g) + (f, g) = (f, (\phi^*\phi + 1)g)$. Hence $(\phi^*\phi + 1)$ is Hermitian. By Theorem 13.25, $(\phi^*\phi + 1)^{-1}$ is Hermitian. But $D((\phi^*\phi + 1)^{-1}) = S$. By Theorem 13.18, $(\phi^*\phi + 1)^{-1}$ is s.a. and s.v., and by Theorem 13.25, $(\phi^*\phi + 1)$ is s.a. It is readily seen that $(P + 1)^* = P^* + 1$. Hence if $(P + 1)^* = (P + 1)$, then $P^* = P$. Therefore $\phi^*\phi$ is s.a., and hence linear and closed; since ϕ and ϕ^* are s.v., $\phi^*\phi$ is s.v.; by Corollary 1 of Theorem 13.14, $D(\phi^*\phi)$ is dense in S. The rest of the theorem follows by a similar argument in which it is shown, as above, that $(\phi\phi^* + 1)$ is Hermitian, that $\phi\phi^*$ is s.a., and that $\phi\phi^*$ has the other properties stated.

THEOREM 13.27. *If ϕ is linear, closed, s.v., with $D(\phi)$ dense in S, then $|||(\phi^*\phi + 1)^{-1}||| \leq 1$ and $|||\phi(\phi^*\phi + 1)^{-1}||| \leq 1$; furthermore, $(\phi^*\phi + 1)^{-1}$ and $\phi(\phi^*\phi + 1)^{-1}$ are closed, bounded, and defined over all S.*

Proof: It was shown in the proof of the preceding theorem that $(\phi^*\phi + 1)^{-1}$ is s.v. and that its domain is S. Let g be any element of S and let $f = (\phi^*\phi + 1)^{-1}g$. Therefore $f \in D(\phi^*\phi + 1)$, and, a fortiori, $f \in D(\phi)$. Thus ϕf has a unique value, that is, $\phi(\phi^*\phi + 1)^{-1}g$ has a unique value for every g in S. By Theorem 13.23, ϕ^* is s.v., so that $\phi^*\phi + 1$ is s.v. Hence $(\phi^*\phi + 1)f$ has the unique value g. Now $((\phi^*\phi + 1)f, f) = (\phi^*\phi f, f) + (f, f) = (\phi f, \phi f) + (f, f) = \|\phi f\|^2 + \|f\|^2$. If f is replaced by $(\phi^*\phi + 1)^{-1}g$, then this relation becomes $(g, (\phi^*\phi + 1)^{-1}g) = \|\phi(\phi^*\phi + 1)^{-1}g\|^2 + \|(\phi^*\phi + 1)^{-1}g\|^2$. By Schwarz's Lemma, $\|g\| \cdot \|(\phi^*\phi + 1)^{-1}g\| \geqq \|(\phi^*\phi + 1)^{-1}g\|^2 + \|\phi(\phi^*\phi + 1)^{-1}g\|^2$. Since the last term is non-negative, it follows that $\|g\| \geqq \|(\phi^*\phi + 1)^{-1}g\|$, and since the first term of the right member is non-negative, $\|g\|^2 \geqq \|\phi(\phi^*\phi + 1)^{-1}g\|^2$, that is, $\|g\| \geqq \|\phi(\phi^*\phi + 1)^{-1}g\|$. Hence $|||(\phi^*\phi + 1)^{-1}||| \leqq 1$ and $|||\phi(\phi^*\phi + 1)^{-1}||| \leqq 1$. By Theorem 13.25, $(\phi^*\phi + 1)^{-1}$ and $\phi(\phi^*\phi + 1)^{-1}$ are linear and closed, and these operators have already been shown to be bounded and defined over all S, so that the proof is complete.

The operator $(\phi^*\phi + 1)^{-1}$ has been shown to be Hermitian. Let $(\phi^*\phi + 1)^{-1}g = f$, where $g \in S$. Then $(g, (\phi^*\phi + 1)^{-1}g) = ((\phi^*\phi + 1)f, f) = \|\phi f\|^2 + \|f\|^2 \geqq 0$, and $(g_1(\phi^*\phi + 1)^{-1}h) = 0$ implies $(\phi^*\phi + 1)^{-1}g = f = 0$, $g = 0$. Hence $(\phi^*\phi + 1)^{-1}$ is definite and $0 \leqq \underline{|||}(\phi^*\phi + 1)^{-1}\underline{|||} \leqq \overline{|||}(\phi^*\phi + 1)^{-1}\overline{|||} \leqq 1$.

THEOREM 13.28. *Let ϕ be as in Theorem 13.26. It is obvious that* $D(\phi^*\phi) \subset D(\phi)$. *Let ϕ_1 be such that $D(\phi_1) = D(\phi^*\phi)$ and such that, if $f \in D(\phi_1)$, $\phi_1 f = \phi f$. Then $\tilde{\phi}_1 = \phi$.*

Proof: Since $\phi_1 \subset \phi$ and ϕ is closed and linear, $\tilde{\phi}_1 \subset \phi$. Hence $[G(\phi_1)] \subset G(\phi)$. It must be shown that $[G(\phi_1)] = G(\phi)$. (Remark: if M and N are c.l.m. with $M \subset N$, then any element $f \in N$ can be represented as $f = f_1 + f_2$, where $f_1 \in M$ and $f_2 \in \ominus M$. Since $f \in N$ and $f_1 \in M \subset N$, it follows that $f_2 = f - f$ is in N. Therefore $f_2 \in (\ominus M)N$. If it can be shown that $f_2 = 0$ for any $f \in N$, then $M = N$. But it can be shown that $f_2 = 0$ by showing that $(\ominus M)N = 0$.) Suppose $K \epsilon \ominus [G(\phi_1)] \cdot G(\phi)$. Since $K \epsilon G(\phi)$, it may be represented by $\langle k, \phi k \rangle$. As in the proof of Theorem 13.26, any element F of the form $\langle f, 0 \rangle$ may be represented as $\langle g, \phi g \rangle + \langle -\phi^* h, h \rangle$, where $h = -\phi g$. Since $h \epsilon D(\phi^*)$, $\phi^*\phi g$ is defined, and $g \epsilon D(\phi^*\phi) = D(\phi_1)$ so that $\phi g = \phi_1 g$. Hence $\langle g, \phi g \rangle \epsilon G(\phi_1) \subset [G(\phi_1)]$ and $K \perp \langle g, \phi g \rangle$. Since $\langle -\phi^* h, h \rangle \epsilon \ominus G(\phi)$, $K \perp \langle -\phi^* h, h \rangle$. Hence K is orthogonal to an arbitrary element $F = \langle f, 0 \rangle$ and $(\langle k, \phi k \rangle, \langle f, 0 \rangle) = 0$, so that $(k, f) = 0$ for all f. Thus $k = 0$. Since ϕ is s.v., $\phi k = 0$. This completes the proof.

THEOREM 13.29. *If ϕ is a semi-definite linear Hermitian operator, then $|(\phi f, g)| \leqq \sqrt{(\phi f, f)(\phi g, g)}$ for every f and g in $D(\phi)$.*

Proof: Since ϕ is semi-definite, $(\phi(f - g), f - g) \geqq 0$ for f and g in $D(\phi)$. Hence $2\Re(\phi f, g) \leqq (\phi f, f) + (\phi g, g)$. If the argument in the proof of Schwarz's Lemma is applied to this relation, it follows that $|(\phi f, g)| \leqq$ $\leqq \sqrt{(\phi f, f)(\phi g, g)}$, $(\phi p, p)$ being $\geqq 0$ for $p \epsilon D(\phi)$.

If ϕ is as in the preceding theorem, and if f is such that $(\phi f, f) = 0$, then $(\phi f, g) = 0$ for all $g \epsilon D(\phi)$. If $D(\phi)$ is dense in S, then, by continuity, $(\phi f, g) = 0$ for all g in S. Therefore $\phi f = 0$, that is, $\phi^{-1}(0) = f$. By Theorem 13.25, ϕ^{-1} is linear. If ϕ^{-1} is s.v., then $f = 0$ and ϕ **is** definite. This proves the following

Corollary: *If ϕ is a semi-definite linear Hermitian operator with $D(\phi)$ dense in S, and if ϕ^{-1} is s.v., then ϕ is definite.*

Let A and A^* be linear, closed, s.v., with domains dense in S. Let $N(A)$ be the set of elements f of $D(A)$ such that $Af = 0$; let $N(A^*)$ be defined analogously. By the second remark after Definition 13.13 it follows that $N(A^*)$ consists of all elements f such that $(f, Ag) = (0, g) = 0$ for all $g \epsilon D(A)$, that is, of all elements f orthogonal to $R(A)$. Hence $N(A^*) = \ominus R(A)$. Similarly, $N(A) = \ominus R(A^*)$.

This proves

THEOREM 13.30: *In the above notation*, $N(A) = \ominus R(A^*)$, $N(A^*) = \ominus R(A)$, *and if* $A = A^*$, *then* $N(A) = \ominus R(A)$.

Appendix II.

Let A be a linear closed operator in S. Then $G(A)$ is a c.l.m. in $S \times S$, and determines a projection $\mathcal{E}_A$ in $S \times S$. (In this discussion an operator in $S \times S$ is denoted by a script letter, while an operator in S is denoted by a Latin letter.) It is desirable to represent $\mathcal{E}_A$ by operators in S.

If $\mathcal{A}$ is an arbitrary operator in $S \times S$, then $\mathcal{A}\langle f, g\rangle = \langle h, k\rangle$, where h and k are each functions of f and g. But if $\mathcal{A}$ is linear, then $\mathcal{A}\langle f, g\rangle = \mathcal{A}\langle f, 0\rangle + \mathcal{A}\langle 0, g\rangle = \langle A_{11}f, A_{21}f\rangle + \langle A_{12}g, A_{22}g\rangle = \langle A_{11}f + A_{12}g, A_{21}f + A_{22}g\rangle$, where $A_{ij}(i, j = 1, 2)$ are operators in S. Thus $\mathcal{A}$ may be represented by the matrix $\begin{Vmatrix} A_{11} & A_{12} \\ A_{21} & A_{22} \end{Vmatrix}$. It is apparent that all A_{ij} are linear; if $\mathcal{A}$ is closed, continuous, s.v., or bounded, then all A_{ij} have the same property.

Now suppose that $\mathcal{A}$ is a projection $P_{G(A)}$, where A is linear and closed. Then $P_{G(A)}\langle f, g\rangle = \langle h, Ah\rangle = \langle P_{11}f + P_{12}g, P_{21}f + P_{22}g\rangle$, where P_{ij} denotes $(P_{G(A)})_{ij}$. Thus $P_{21} = AP_{11}$ and $P_{22} = AP_{12}$. But $P_{G(A)}\langle f, 0\rangle = \langle P_{11}f, P_{21}f\rangle$. In the proof of Theorem 13.26 it was shown that, if $P_{G(A)}\langle f, 0\rangle = \langle h, Ah\rangle$, then $f = (A^*A + 1)h$, that is, $h = (A^*A + 1)^{-1}f$. But $h = P_{11}f$; therefore $P_{11} = (A^*A + 1)^{-1}$ and $P_{21} = A(A^*A + 1)^{-1}$. By Theorem 13.27, P_{11} and P_{21} are bounded and defined over S. By Definition 13.13, $G(A^*) = \ominus \bar{U}G(A)$. Since $P_{\bar{U}G(A)} = \bar{U}P_{G(A)}\bar{U}^{-1}$, $P_{G(A^*)} = 1 - \bar{U}P_{G(A)}\bar{U}^{-1}$. Therefore $\bar{U}^{-1}\langle f, g\rangle = \langle g, -f\rangle$, $P_{G(A)}\bar{U}^{-1}\langle f, g\rangle = \langle P_{11}g - P_{12}f, P_{21}g - P_{22}f\rangle$, $\bar{U}P_{G(A)}\bar{U}^{-1}\langle f, g\rangle = \langle P_{22}f - P_{21}g, -P_{12}f + P_{22}g\rangle$, and $P_{G(A^*)}\langle f, g\rangle = \langle (1 - P_{22})f + P_{21}g, P_{12}f + (1 - P_{11})g\rangle$. Thus $P^*_{11} = 1 - P_{22}$, $P^*_{12} = P_{21}$,

$P^*_{21} = P_{12}$, and $P^*_{22} = 1 - P_{11}$, where P^*_{ij} denotes $(P_{G(A^*)})_{ij}$. Since $A^{**} = A$, $P^*_{21} = A^*(AA^* + 1)^{-1} = P_{12}$, and $P_{22} = AA^*(AA^* + 1)^{-1}$. By Theorem 13.27, P_{12} is bounded and defined over S; it is apparent that P_{22} must also have these properties. It is convenient to tabulate these results.

$$P_{11} = (A^*A + 1)^{-1} \qquad P_{12} = A^*(AA^* + 1)^{-1}$$

$$P_{21} = A(A^*A + 1)^{-1} \qquad P_{22} = AA^*(AA^* + 1)^{-1}$$

CHAPTER XIV.

COMMUTATIVITY, REDUCIBILITY

In the preceding chapter an important class of operators was considered: the projections. The next task will be to define and (partially) analyze another important class: the unitary operators.

Definition 14.1. *An isomorphism of S is a biunique mapping $f \rightarrow Uf$ of S upon itself which leaves invariant all formal relations used in the postulational characterization of S.* (This accords with general usage.) *All these formal relations may be expressed in terms of the operations* af, $f + g$, (f, g). *Hence it is postulated of a unitary transformation that, for arbitrary* a, f, g,

1) $U(af) = aUf$, 2) $U(f+g) = Uf + Ug$, 3) $(Uf,Ug) = (f, g)$.

It is evident that such an isomorphism $f \rightarrow Uf$ may be looked upon as an operator U. Whenever it is desired to emphasize the operatorial character of an isomorphism (and this will usually be the case) *it will be called a unitary operator.*

It is necessary to introduce some direct operatorial characterizations that a transformation be unitary.

THEOREM 14.1. *An operator U is unitary if and only if*

1) *U is s.v., linear, and closed,*

2) $\|Uf\| = \|f\|$ *for all elements f in S,*

3) $D(U) = R(U) = S$.

The third condition may be replaced by the following (which is apparently weaker):

3') D(U) <u>and</u> R(U) <u>are dense in</u> S.

Proof: It will be shown that 1, 2, and 3 are necessary and that 1, 2, and 3' are sufficient.

Necessity of the conditions: condition 3 follows from the fact that the transformation $f \to Uf$ maps S on itself; condition 2 follows from part 3 of Definition 14.1 when f = g; it is evident from Definition 14.1 that U is s.v. and linear, and closure follows from continuity over $D(U) = S(\|Uf - Ug\| = \|f - g\|)$

Sufficiency of the conditions: from 1 (linearity) and 2 it follows that $\|Uf - Ug\| = \|f - g\|$. Thus if Uf = Ug, then f = g, so that U has a s.v. inverse U^{-1} which (evidently) satisfies conditions 1, 2, 3'. Since $\|Uf - Ug\| = \|f - g\|$, U is continuous over D(U); by Theorem 13.11, D(U) is closed; since D(U) is dense in S, D(U) = S. As the same argument applies to U^{-1}, $D(U^{-1}) = S$, and U is a biunique mapping of S on itself. Since U is linear, conditions 1 and 2 of Definition 14.1 obtain. It remains to show that (Uf, Ug) = (f, g). By 2, this condition holds for f = g. In particular, it follows from the condition $(U\frac{f+g}{2}, U\frac{f+g}{2}) - (U\frac{f-g}{2}, U\frac{f-g}{2}) = (\frac{f+g}{2}, \frac{f+g}{2}) - (\frac{f-g}{2}, \frac{f-g}{2})$ that $\mathcal{R}(Uf, Ug) = \mathcal{R}(f, g)$. If g is replaced by ig, this argument shows that $\mathcal{J}(Uf, Ug) = (f, g)$. This completes the proof.

<u>Remark 1.</u> The preceding theorem shows that every unitary operator is bounded.

<u>Remark 2.</u> It is evident from condition 3 of Definition 14.1 that condition 2 of the preceding theorem may be replaced by the (apparently) stronger condition (Uf, Ug) = (f, g). By Theorem 13.21, $D(U^*) = S$, so that this condition may be written in the form $(U^*Uf, g) = (f, g)$; since D(U) = S, this condition implies that $U^*U = 1$. As the condition $D(U^*U) = S$ implies the condition D(U) = S, it follows that another set of necessary and sufficient conditions

that U be unitary is that

1) U is s.v., linear, and closed,

2) $U^*U = 1$

3) R(U) is dense in S.

Remark 3. If U is unitary, then $U^*Uf = f$, $UU^*Uf = Uf$, so that $UU^*g = g$ if $g = Uf$, that is, if $g \varepsilon T(U)$. But UU^* is continuous and R(U) is dense in S, so that $UU^*g = g$ for all elements g. Hence $UU^* = 1$. But, from Remark 2, $U^*U = 1$. Thus

$$UU^* = U^*U = 1.$$

Conversely, this relation (regardless of the unitary character of U) is equivalent to the condition $U^{-1} = U^*$. Hence it implies that $R(U) = D(U^{-1}) = D(U^*) = S$. Thus condition 2 and 3 of the preceding remark may be replaced by the condition

$\bar{2}$) $U^*U = UU^* = 1$

or by the condition

$\bar{\bar{2}}$) $U^{-1} = U^*$.

Remark 4. The argument used at the end of the proof of Theorem 14.1 and in Remark 2 may be used to show that condition $\bar{2}$ of Remark 3 (that is, conditions 2 and 3 of Remark 2) may be replaced by the condition

2^*) $\|Uf\| = \|U^*f\| = \|f\|$ for all elements f in S.

Those unitary operators which are also Hermitian are of some interest per se. They are bounded (Remark 1), s.v., linear, closed, and their Hermitian character is expressed by the condition $U = U^*$, that is, $U^2 = 1$ (condition $\bar{\bar{2}}$ of Remark 3). Conversely, the condition $U^2 = 1$ implies that $D(U) = S$, so that the property of being Hermitian implies the property of being s.a., s.v. (Theorem 13.18), and therefore also linear and closed. This leads to

THEOREM 14.2. *The set of unitary operators U which are also Hermitian is exactly the set of Hermitian operators such that* $U^2 = 1$.

(Proof above.)

Remark. Since $\left(\frac{U+1}{2}\right)^2 = \frac{U^2+2U+1}{4} = \frac{U+(\frac{1}{2}+\frac{1}{2}U^2)}{2}$, the condition $U^2 = 1$ is equivalent to the condition $\left(\frac{U+1}{2}\right)^2 = \frac{U+1}{2}$. Hence the correlation

$$\frac{U+1}{2} = E, \qquad 2E - 1 = U$$

sets up a biunique correspondence between the set of all unitary-Hermitian operators U and the set of all projections E.

Since unitary operators were defined as isomorphisms of S upon itself, the set of all unitary operators forms a group. Hence

1) 1 is unitary,

2) U^{-1} is unitary along with U,

3) UV is unitary along with U and V.

(It is evident that these propositions may be readily verified by means of any other one of the above characterizations of the property of being unitary.)

If U and V are unitary-Hermitian, then UV will be unitary-Hermitian if UV = VU (since UVf, g) = (f, VUg), so that $(UV)^* = (VU)$. Let the projections E and F correspond respectively to U and V; let $E = P_M$ and $F = P_N$; assume that UV = VU; then $G = \frac{UV+1}{2} = \frac{(2E-1)(2F-1)+1}{2} = 1 - E - F + 2EF = 1 - \{E(1-F) + F(1-E)\}$ $= EF + (1-E)(1-F)$ corresponds to UV. Since UV = VU, EF = FE, and M and N commute. But

$$E(1-F) + F(1-E) = P_{[M(\ominus N),\ N(\ominus M)]},$$

$$EF + (1-E)(1-F) = P_{[MN,\ (\ominus M)(\ominus N)]},$$

so that

$$G = P_{\ominus [M(\ominus N),\ N(\ominus M)]} = P_{[MN,\ (\ominus M)(\ominus N)]}.$$

It is now desirable to introduce the concept of commutativity. If **A** and B are two s.v. operators, it may well happen that D(AB) and D(BA) are distinct or even empty. It is a case of particular interest when BAf and ABf exist and are equal for every element f in the domain of one of the operators, say A. For BAf to exist for all f ε D(A) it is necessary and sufficient that D(B) $\supset$ R(A). For ABf to exist for all f ε D(A) it is necessary and sufficient that D(B) $\supset$ D(A) and that Bf ε D(A) for all f ε D(A). If under these conditions BAf = ABf for every f ε D(A), it is apparent that it cannot be said that AB = BA, for these two operators may have distinct domains; all that can be said is that AB $\supset$ BA.

Because of the conditions on D(B) it is convenient to assume that D(B) = S. In this case it is possible to assert (conversely to the preceding remark) that if AB $\supset$ BA, then ABf = BAf for all f ε D(A). But it would introduce an undesirable element of asymmetry between the roles of A and B to say on this basis that A and B commute. Symmetry is retained between A and B in

<u>Definition 14.2.</u> <u>Let</u> A <u>and</u> B <u>be two s.v.operators.</u> A <u>and</u> B <u>are said to commute if either</u> 1) D(A) = S <u>and</u> BA $\supset$ AB (<u>so that for</u> f ε D(B) <u>it follows that</u> Af ε D(B) <u>and</u> ABf = BAf), <u>or</u> 2) D(B) = S <u>and</u> AB $\supset$ BA (<u>so that for</u> f ε D(A) <u>it follows that</u> Bf ε D(A) <u>and</u> ABf = BAf).

<u>Remark.</u> If D(A) = D(B) = S, then D(AB) = D(BA) = S and either of the conditions in the preceding definition implies that AB = BA, this being the customary definition of commutativity. Thus in the present extended sense of the word, the condition BA $\supset$ AB may also be utilized when D(A) = S regardless of the nature of D(B); and the condition AB $\supset$ BA may always be utilized when D(B) = S, regardless of the nature of D(A).

It should be noted that no definition of commutativity has been given in the case where D(A) $\neq$ S and D(B) $\neq$ S. This case is at present insufficiently analyzed, though in certain special cases (to be discussed later) satisfactory definitions can be given.

THEOREM 14.3. *If A is a s.v. operator, it follows that*

1) *A commutes with 1.*

2) *If A0 = 0, then A commutes with 0.*

3) *If D(A) = S, then A commutes with A.*

4) *If A commutes with B, if* B^{-1} *is s.v. (so that Bf = Bg only when f = g), and if either D(A) = S or* $D(B^{-1}) = R(B) = S$, *then A commutes with* B^{-1}.

5) *If A commutes with B and C, then A commutes with BC.*

6) *If A is linear and commutes with B and C, then A commutes with aB and B ± C.*

7) *If A commutes with* $B_1, B_2, \ldots$, *if* $B = \lim_{n\to\infty} B_n$ (i.e., Bf is defined if and only if all $B_n f$ are defined and $\lim_{n\to\infty} B_n f$ exists, and in this event $Bf = \lim_{n\to\infty} B_n f$), *and if either D(A) = S and A is continuous or D(B) = S and A is closed, then A commutes with B.*

8) *If A commutes with B, if* A^* *and* B^* *are s.v., and if either* $D(A) = D(A^*) = S$ *or* $D(B) = D(B^*) = S$, *then* A^* *commutes with* B^*.

9) *If A is linear, continuous, and commutes with B, and if* $\tilde{B}$ *is s.v., then A commutes with* $\tilde{B}$.

Proof: Parts 1 and 2 are evident since D(1) = D(0) = S. Parts 3 to 7 are readily verified from Definition 14.2; in each of these parts the case where D(A) = S (requiring the use of condition 1 of Definition 14.2) must be considered separately from the case where D(A) ≠ S (requiring the use of condition 2; note the remark). Part 9 follows at once if D(A) = S (condition 1); but if D(A) ≠ S, then D(B) = S and $B = \tilde{B}$, so that the remainder of part 9 is immediate. It remains to consider part 8, where, by reason of symmetry, it may be assumed that $D(B) = D(B^*) = S$. Let g be an element of $D(A^*)$. For arbitrary f ε D(A) it follows that $(f, B^*A^*g) = (Bf, A^*g) = (ABf, g) = (BAf, g) = (Af, B^*g)$. Hence there exists an element $g^* = B^*A^*g$ such that $(f, g^*) = (Af, B^*g)$ for all f ε D(A).

But this last condition implies that $B^*g \in D(A^*)$ and $A^*B^*g = B^*A^*g$. Hence $A^*B^* \supset B^*A^*$.

A refinement of the concept of commutativity is given by

Definition 14.3. *Let A and B be two s.v. operators. A and B are said to commute adjointly (c.a.) if either 1) $D(A) = D(A^*) = S$ while A and A^* commute with B, or 2) $D(B) = D(B^*) = S$ while B and B^* commute with A.*

Remark. If $D(A) = D(A^*) = S$, then, by Theorems 13.22 (applied to A^*) and 13.21, both A and A^* are bounded; if $D(B) = D(B^*) = S$, then B and B^* are bounded. If both of these conditions obtain, then, by condition 1 of the preceding definition, $AB = BA$, $A^*B = BA^*$, and by condition 2, $AB = BA$, $AB^* = B^*A$. The first equations of these pairs are the same, the second equations arise from each other by applying the operation *. (It is evident that $(XY)^* = Y^*X^*$ for bounded operators X, Y). Thus condition 1 of the preceding definition may always be utilized when $D(A) = D(A^*) = S$ regardless of the nature of D(B) and $D(B^*)$; and condition 2 may always be utilized when $D(B) = D(B^*) = S$, regardless of the nature of D(A).

The present definition of c.a. does not apply in the case where $D(A) \neq S$ and $D(B) \neq S$. We will see later that in these cases too a satisfactory definition of c.a. can be given. Thus c.a. will turn out to be a more natural notion than commutativity itself. In this aspect, Theorem 14.5 and the remark which precedes Definition 14.4, are quite instructive.

THEOREM 14.4. *If A is a s.v. operator, it follows that*

1) *A c.a. with 1.*
2) *If $A0 = 0$, then A c.a. with 0.*
3) *If $D(A) = D(A^*) = S$, then A c.a. with A.*
4) *If A c.a. with B, if B^{-1} is s.v.* (see Theorem 14.3, part 4), *and if either $D(A) = D(A^*) = S$ or $D(B^{-1}) = R(B) = D(B^{*-1}) = R(B^*) = S$, then A c.a. with B^{-1}.*

5) <u>If</u> A <u>c.a.</u> <u>with</u> B <u>and</u> C, <u>then</u> A <u>c.a.</u> <u>with</u> BC.

6) <u>If</u> A <u>is linear and c.a. with</u> B <u>and</u> C, <u>then</u> A <u>c.a.</u> <u>with</u> aB <u>and</u> $B \pm C$.

7) <u>If</u> A <u>c.a.</u> <u>with</u> $B_1, B_2, \ldots$, <u>if</u> $B = \lim_{n \to \infty} B_n$ (see Theorem 14.3, part 7), <u>and</u> <u>if either</u> $D(A) = D(A^*) = S$ <u>or</u> $D(B) = D(B^*) = S$ <u>and</u> A <u>is closed</u>, <u>then</u> A <u>c.a.</u> <u>with</u> B.

8) <u>If</u> A <u>c.a.</u> <u>with</u> B, <u>and if</u> B^* <u>is s.v.</u>, <u>then</u> A <u>c.a.</u> <u>with</u> B^*.

9) <u>If</u> A <u>is linear, continuous, and c.a. with</u> B, <u>and if</u> $\tilde{B}$ <u>is s.v.</u>, <u>then</u> A <u>c.a.</u> <u>with</u> $\tilde{B}$.

<u>Remark</u>: This theorem is obviously analogous to Theorem 14.3. However, part 8 is stronger here then there; two applications of part 8 (once to B and then to A) lead to part 8 of Theorem 14.3; the direct analogue of part 8 does not hold for commutativity itself. In fact, this is the principal reason for the introduction of the terminology "commutes adjointly".

Proof: The proofs of parts 1 to 7 are the same as in Theorem 14.3. In particular, the condition $D(A) = D(A^*) = S$ in part 7 implies that A is bounded and therefore continuous. In part 9 it may again be assumed that $D(A) = D(A^*) = S$; since the condition $D(B) = S$ implies that $B = \tilde{B}$, the analogy is complete. It remains to consider part 8. If $D(B) = D(B^*) = S$, then B is bounded, $\tilde{B}$ is s.v., and $B = \tilde{B} = B^{**}$. If B is replaced by B^*, then in Definition 14.3 the operators B and B^* are replaced by B^* and $B^{**} = B$, that is, no change is made. If $D(A) = D(A^*) = S$, then $A = \tilde{A} = A^{**}$, and it must be shown that if B commutes with A and A^*, then B^* commutes with A and A^*, that is, with A^* and $A^{**} = A$. But this follows directly from Theorem 14.3, part 9.

It is of particular interest to discuss the situation in which A c.a. with an operator which is either a projection or is unitary.

<u>THEOREM 14.5. An operator</u> A <u>c.a. with a unitary operator</u> U <u>if and only if</u> A <u>is invariant under</u> U, <u>that is</u>, $A = UAU^{-1}$.

Proof: Right multiplication of $A = UAU^{-1}$ with U gives $AU = UA$; left multiplication with U^{-1} gives $U^{-1}A = AU^{-1}$, and as $U^{-1} = U^*$, $U^*A = AU^*$. Thus the condition is sufficient. Conversely, if A commutes with U, $AU \supset UA$, and in particular $f \in D(A)$ implies $Uf \in D(A)$. If A c.a. with U, then this assertion holds also for $U^* = U^{-1}$. It follows that D(A) is transformed by both U and U^{-1} into part of itself; therefore it is invariant under U, and so A, and UAU^{-1} have the same domain. Since $AU \supset UA$, $A \supset UAU^{-1}$, this gives $A = UAU^{-1}$.

If E is a projection, then $E = E^*$, $D(E) = D(E^*) = S$, and the following three conditions are equivalent: A c.a. with E, A commutes with E, and $AE \supset EA$.

Definition 14.4. If A c.a. with $E = P_M$ (note the preceding remark), then E and M are each said to reduce A.

THEOREM 14.6. *If A and A^* are s.v. with $D(A^*)$ dense in S, then the following four relations together constitute a necessary and sufficient condition that A be reduced by $E = P_M$:*

1) $Ef \in D(A)$ *for* $f \in D(A)$,

2) $Ef \in D(A^*)$ *for* $f \in D(A^*)$,

3) $Af \in M$ *for* $f \in M \cdot D(A)$,

4) $A^*f \in M$ *for* $f \in M \cdot D(A^*)$.

Proof of necessity. Condition 1 is immediate. Condition 2 follows directly from Theorem 14.4, part 8 (where A and B are replaced by E and A). If $f \in M \cdot D(A)$, then $Ef = f$ and $Af = AEf = EAf$, so that $Af \in M$. Condition 4 follows in an analogous manner.

Proof of sufficiency: If $f \in D(A)$, then (by 1) Ef is in D(A), M, and $M \cdot D(A)$. By 3, $AEf \in M$, so that $E(AEf) = AEf$. Similarly, if $g \in D(A^*)$, then $E(A^*Eg) = A^*Eg$. Since $E^* = E$, $(AEf, g) = (EAEf, g) = (AEf, Eg) = (Ef, A^*Eg) =$

$= (f, EA^*Eg) = (f, A^*Eg) = (Af, Eg) = (EAf, g)$. Therefore $(AEf, g) = (EAf, g)$ for all f in D(A) and all g in $D(A^*)$. Since $D(A^*)$ is dense in S, this relation holds for all g. Hence $AEf = EAf$ for all $f \in D(A)$, so that $AE \supset EA$.

Remark: In the preceding theorem the condition that $D(A^*)$ be dense in S may be replaced by the condition that D(A) be dense in S.

Corollary: *If A and A^* are s.v. and defined over all of S, then a necessary and sufficient condition that A be reduced by $E = P_M$ is that $Af \in M$ and $A^*f \in M$ for $f \in M$.*

Proof: This corollary is merely a special case of Theorem 14.6 in which conditions 1 and 2 are necessarily satisfied and in which conditions 3 and 4 reduce to the conditions stated.

Theorem 14.6 and this corollary show that A and A^* may be regarded as operators in merely the space M. Thus the behavior of A may be analyzed by means of subspaces M reducing A.

THEOREM 14.7. *If A is a linear, closed, and s.v. operator, then*

1) *A is reduced by $E = P_M$ if and only if it is reduced by $1 - E = P_{\ominus M}$.*
2) *If A is reduced by every $E_\alpha = P_{M_\alpha}$ where α ranges over a set I (of indices α), then A is reduced by*
   a) $P_{\mathfrak{M}}$, *where* $\mathfrak{M} = \prod_{\alpha \in I} M_\alpha$.
   b) $P_{\mathfrak{N}}$, *where* $\mathfrak{N} = [\ldots, M_\alpha, \ldots]$ *(where α in $[\ldots]$ ranges over all of I).*

Proof: Part 1. By Theorem 14.4, parts 1 and 6, if A c.a. with E, then it c.a. with 1 - E, and if A c.a. with 1 - E, then it c.a. with $1 - (1-E) = E$.

Part 2b. This follows directly from parts 1 and 2a since

$$[\ldots, M_\alpha, \ldots] = \ominus \left(\prod_{\alpha \in I} (\ominus M_\alpha) \right)$$

Part 2a. Case 1. I contains only two elements, say 1 and 2. Since

A c.a. with E_1 and E_2, it follows by induction from Theorem 14.4, part 5, that A c.a. with each of the operators E_1, E_2E_1, $E_1E_2E_1$, ... By Theorem 13.7, this sequence has for limit the projection $P_{M_1 \cdot M_2}$. By Theorem 14.4, part 7, A c.a. with $P_{M_1 \cdot M_2}$.

Case 2. I is a finite set, say (1,...,n). This case follows immediately by induction.

Case 3. I is a countably infinite set, say (1,2,...). Let $\mathfrak{M}_k = \sum_{\alpha=1}^{k} M_\alpha$. Then $P_{\mathfrak{M}_1} \geqq P_{\mathfrak{M}_2} \geqq \ldots$ is a sequence of projections which, by Theorem 13.10, has the limit $P_{\mathfrak{M}}$, where $\mathfrak{M} = \sum_{\alpha=1}^{\infty} M_\alpha$. By Case 2, A c.a. with every $P_{\mathfrak{M}_\beta}$, so that, by Theorem 14.4, part 7, A c.a. with $P_{\mathfrak{M}}$.

Case 4. I is a non-countable set. Let all possible (finite or infinite)sequences M_{α_1}, M_{α_2}, ... be formed from the set of sets M_α. Let f be an element of $\mathfrak{H}$ and let γ be the g.l.b. of all the numbers $\| P_{M_{\alpha_1} \cdot M_{\alpha_2} \cdot \ldots} f \|$ corresponding to all the sequences M_{α_1}, M_{α_2}, ... It is apparent that $\gamma \geqq 0$. Let n be a positive integer. There exists a sequence $M_{\alpha_1^n}$, $M_{\alpha_2^n}$, ... such that $\| P_{M_{\alpha_1^n} \cdot M_{\alpha_2^n} \ldots} f \| < \gamma + \frac{1}{n}$. Let β_1, β_2, ... be any sequence in I containing all the indices α_i^n (n, i=1,2, ...), and let $M = M_{\beta_1} \cdot M_{\beta_2} \cdot \ldots$. Then $M = M_{\beta_1} \cdot M_{\beta_2} \cdot \ldots \subset M_{\alpha_1^n} \cdot M_{\alpha_2^n} \ldots$, $\| P_M f \| \leqq \| P_{M_{\alpha_1^n} \cdot M_{\alpha_2^n} \ldots} f \| < \gamma + \frac{1}{n}$. (Here, and in what follows, use is made of Theorem 13.8 and the discussion of the relation $E \leqq F$ preceding it.) But $\| P_M f \| \geqq \gamma$, and as $\| P_M f \|$ is independent of n, $\| P_M f \| = \gamma$.

It will now be shown that $P_M f = P_{\mathfrak{M}} f$ for all f in S. Let λ be an arbitrary element of I. Since the sequence $\lambda, \beta_1, \beta_2$, ... contains all the indices α_i^n, it follows that $\| P_{M_\lambda \cdot M} f \| = \| P_{M_\lambda \cdot M_{\beta_1} \cdot M_{\beta_2} \ldots} f \| = \gamma$. But $\| P_M f \| = \gamma$, so that $\| P_{M_\lambda \cdot M} f \| = \| P_M f \|$. By the corollary to Theorem 13.8, this relation implies that $P_{M_\lambda \cdot M} f = P_M f$ and, as $P_{M_\lambda \cdot M} f \subset M_\lambda \cdot M \subset M_\lambda$, $P_M f \in M_\lambda$. But this condition holds for each λ in I, so that $P_M f \in \mathfrak{M}$ and $P_{\mathfrak{M}} P_M f = P_M f$. But $\mathfrak{M} \subset M$

and $P_{\mathfrak{M}} \leq P_M$. Hence $P_{\mathfrak{M}} P_M = P_{\mathfrak{M}}$, and therefore $P_{\mathfrak{M}} f = P_M f$.

Let f and f^* be two elements of S, let α_i^n and α_i^{*n} be the respective sets of indices determined as above, and let $\beta_1, \beta_2, \ldots$ be a sequence consisting of all the indices α_i^n and α_i^{*n}. Then $P_M f = P_{\mathfrak{M}} f$ and $P_M f^* = P_{\mathfrak{M}} f^*$. Now let f be restricted to D(A) and let $f^* = Af$. The preceding relations assume the form $P_M f = P_{\mathfrak{M}} f$ and $P_M Af = P_{\mathfrak{M}} Af$. By Case 3 and Theorem 14.6, $P_M f \in D(A)$ and $AP_M f = P_M Af$; hence $P_{\mathfrak{M}} f \in D(A)$ and $AP_{\mathfrak{M}} f = P_{\mathfrak{M}} Af$, where f is an arbitrary element of D(A). Therefore $AP_{\mathfrak{M}} \supset P_{\mathfrak{M}} A$, A commutes with $P_{\mathfrak{M}}$, and (by the remark preceding Definition 14.4) A c.a. with $P_{\mathfrak{M}}$.

<u>THEOREM 14.8. Let T be a separable subset of S(for example, a sequence), and let $A_1, A_2, \ldots$ be a sequence of bounded operators defined over all of S. There exists a separable c.l.m. M containing T and reducing each of the operators $A_1, A_2, \ldots$. (If S is itself separable, then M = S and the theorem is trivial.)</u>

Proof: Let $f_1, f_2, \ldots$ be a sequence of elements of T dense in T. If a c.l.m. M is found which contains $f_1, f_2, \ldots$, then $M \supset T$ since M is closed. Hence it is sufficient to consider the sequence $f_1, f_2, \ldots$.

The set Z of expressions $X^{(1)} \cdot \ldots \cdot X^{(p)} f_n$ is countable, where $n = 1, 2, \ldots, p = 0, 1, 2, \ldots$, where X^j $(j = 1, \ldots, p)$ is an operator A_i or an operator A_i^* $(i = 1, 2, \ldots)$, and where $X^{(1)} \cdot \ldots \cdot X^{(p)}$ represents the operator 1 if $p = 0$. It will be shown that [Z] satisfies the conditions required of M. It is apparent that $f_n \in [Z]$ for each n. By Theorem 12.28, [Z] is separable since the set Z is separable. It remains to show that if $f \in [Z]$, then $A_i f$ and $A_i^* f$ are in [Z] (cf. the corollary of Theorem 14.6). Now if $f \in Z$, then $A_i f$ and $A_i^* f$ are in Z; if $f \in \{Z\}$, then $A_i f$ and $A_i^* f$ are in $\{Z\}$ since A_i and A_i^* are linear; and if $f \in [Z]$, then $A_i f$ and $A_i^* f$ are in

[Z] since A_i and A_i^* are continuous. (Note that closedness would be insufficient.) This completes the proof.

THEOREM 14.9. *Let T be a separable subset of S (for example, a sequence), and let A_1, A_2, ... be a sequence of linear, closed, s.v. operators in S. There exists a separable c.l.m. M containing T and reducing each of the operators A_1, A_2, ...*

Proof: As before, it is sufficient to consider T as being a sequence f_1, f_2, ... The operators of the sequence $\bar{U}$ P_X, P_Y, $P_{G(A_1)}$, $P_{G(A_2)}$, ... are bounded and defined over S × S. (Cf. Definition 13.11 and the remark preceding Definition 13.8). Let [Z] be the set constructed as in the preceding proof with respect to the sequence of points $\langle f_1, 0\rangle$, $\langle f_2, 0\rangle$, ... Then [Z] reduces each of these operators and, by the corollary of Theorem 14.5, if $\langle f, g\rangle \in [Z]$, then $\bar{U}\langle f, g\rangle = \langle -g, f\rangle$, $P_X\langle f,g\rangle = \langle f, 0\rangle$, and $P_Y\langle f,g\rangle = \langle 0, g\rangle$ are all in [Z].

Let M be the set of all elements f such that $\langle f, 0\rangle \in [Z]$ and let N be the set of all elements g such that $\langle 0, g\rangle \in [Z]$. It is apparent that $M = I_X^{-1}([Z]\, X)$; since X is a c.l.m., M is a c.l.m.; since [Z] is separable, M is separable; M obviously contains the elements f_1, f_2, ... It remains to show that M reduces A_1, A_2, ... If $\langle f, g\rangle \in [Z]$, then, by the last assertion of the preceding paragraph, $f \in M$ and $g \in N$. But if $f \in M$, then $\langle f, 0\rangle \in [Z]$, $\langle 0, f\rangle \in [Z]$ (by the operator $\bar{U}$), and $f \in N$. Hence $M \subset N$. It follows similarly that $M \supset N$, so that $M = N$. Therefore, if $\langle f, g\rangle \in [Z]$, then f and g are in M. Again, if $\langle f, g\rangle \in \ominus[Z]$, then $\langle f, g\rangle \perp \langle h, k\rangle$ for every element $\langle h, k\rangle \in [Z]$. Therefore $(f,h) + (g, k) = 0$ for every h and k in M. Since M is linear, $0 \in M$. Hence $(f, h) + (g, 0) = 0$ for all h in M, and $f \in \ominus M$. It follows similarly that $g \in \ominus M$.

Let f be any element of $D(A_i)$. Then $\langle f, A_i f\rangle$ is any element of $G(A_i)$.

Since [Z] reduces $P_{G(A_i)}$, $P_{G(A_i)}P_{[Z]}\langle f, A_i f\rangle = P_{[Z]}P_{G(A_i)}\langle f, A_i f\rangle =$ $= P_{[Z]}\langle f, A_i f\rangle$. Therefore $P_{[Z]}\langle f, A_i f\rangle \in G(A_i)$; let $P_{[Z]}\langle f, A_i f\rangle =$ $= \langle g, A_i g\rangle$. If $\langle f, A_i f\rangle$ is resolved by Theorem 12.23 with respect to [Z], then, since $\langle f, A_i f\rangle$ and one component are in $G(A_i)$, the other component is also. Hence $\langle f, A_i f\rangle = \langle g, A_i g\rangle + \langle h, A_i h\rangle$, where $\langle g, A_i g\rangle \in [Z]$ and $\langle h, A_i h\rangle \in \ominus[Z]$. Therefore $f = g + h$, where, by the preceding paragraph, $g \in M$, $g \in D(A_i)$, and $h \in \ominus M$, $h \in D(A_i)$; likewise $A_i f = A_i g + A_i h$, where $A_i g \in M$ and $A_i h \in \ominus M$. Since M is a c.l.m., it determines a projection P_M. Since $P_M f = g$, $P_M f \in D(A_i)$ and $A_i P_M f = A_i g = P_M A_i f$ (since $A_i g \in M$). As f was an arbitrary element of $D(A_i)$, it follows that $A_i P_M \supset P_M A_i$, so that A_i and P_M commute. This completes the proof.

THEOREM 14.10. *Let $A_1, A_2, \ldots$ be a sequence of linear, closed, s.v. operators in S. There exists a set of separable and non-empty c.l.m.'s M_α (α ranging over a suitable set J (of indices)) such that 1) if $\alpha \neq \beta$, then M_α and M_β are orthogonal, 2) $[\ldots, M_\alpha, \ldots] = S$ (α in $[\ldots]$ ranging over all of J), and 3) each set M_α reduces each operator A_i.*

Proof. The sets M_α will be constructed in such a manner that J will be a set of Cantor's ordinal numbers, namely, the set of all numbers $\alpha < \bar{\alpha}_0$ for a suitable number $\bar{\alpha}_0$. Therefore the sets M_α will be defined by transfinite induction; $\bar{\alpha}_0$(and hence J itself) will be determined only at the end of the process. The transfinite induction is as follows: suppose that, for each $\alpha < \alpha_0$, M_α has been defined so as to satisfy conditions 1 and 3 of the theorem (for each $\alpha < \alpha_0$, M_α is separable, non-empty, and reduces each A_i). Let $S_{\alpha_0} = [\ldots, M_\alpha, \ldots]$, where α ranges over all ordinal numbers less than α_0. If $S_{\alpha_0} = S$, let I be the set of all ordinal numbers α less than α_0. Then condition 2 is also satisfied and the proof is complete. In this case the induction stops at the number α_0.

But if $S_{\alpha_0} \neq S$, then, by Theorem 14.7, each A_i is reduced by $\ominus S_{\alpha_0}$. By Theorem 14.9 there exists a separable and non-empty c.l.m. $M_{\alpha_0} \subset \ominus S_{\alpha_0}$ which reduces each A_i. Thus the sets M_α, $\alpha \leq \alpha_0$, satisfy conditions 1 and 3 and the entire set of sets M_α is defined by transfinite induction.

Since S itself has an ordinal number Ω and since the ordinal number of $[\ldots, M_\alpha, \ldots]$, $\alpha < \alpha_0$, is not less than α_0, the process can reach no number $\alpha_0 > \Omega$. Hence it must stop at some number $\alpha_0 = \bar{\alpha}_0 \leq \Omega$, and when it stops, $S_{\bar{\alpha}_0} = S$. This completes the proof.

Remark. By Theorems 12.26 and 12.27, each set M_α is a finite-dimensional Euclidean space or a Hilbert space. (More accurately, each set M is isomorphic with such a space. But, for the sake of brevity, the words "isomorphic with" will always be omitted in what follows.) If the sets M_α are infinite in number, they may be classified into mutually exclusive, countably infinite classes. The sums of the sets M_α of each such class may themselves be used as sets M_α, and they are all Hilbert spaces. If the number of sets M_α is finite, then S is separable and the theorem can be satisfied with the use of only one set $M_\alpha = S$. In this case $M_\alpha = S$ is a finite-dimensional Euclidean space or a Hilbert space.

Thus it turns out that, unless S is a finite-dimensional Euclidean space, all the sets M_α may be chosen as Hilbert spaces.

Theorem 14.10 shows that any finite or countable finite set of linear, closed, s.v. operators may be simultaneously reduced by a system of separable and non-empty c.l.m's M_α having properties 1 and 2. It is therefore of interest to discuss the nature of such systems M_α and the way operators A reduced by them are determined by their behavior in the individual sets M_α. The following theorems contribute to such a discussion.

It should be remarked that these theorems (particularly Theorem 14.12) will be of use even in separable spaces where it will sometimes be of importance to construct operators A from their contractions in c.l.m.'s M_α (of the sort described above) which reduce them.

It will not be assumed in what follows that the sets M_α are either separable or non-empty.

THEOREM 14.11. Let M_α be a system of c.l.m's, α ranging over a set J (of indices), such that 1) if $\alpha \neq \beta$, then M_α and M_β are orthogonal, and 2) $[\ldots, M_\alpha, \ldots] = S$ (α in $[\ldots]$ ranging over all of J).

Then every element f in S has a unique representation of the form $f = \sum_{\alpha \in J} f_\alpha$, $f_\alpha \in M_\alpha$, where $f_\alpha = 0$ for all $\alpha \in J$ aside from a finite or countable subset, and where the sum of the non-vanishing elements f_α is convergent. Incidentally, $f_\alpha = P_{M_\alpha} f_\alpha$ and $\|f\|^2 = \sum_{\alpha \in J} \|f_\alpha\|^2$.

Conversely, a meaning attaches to each series $\sum_{\alpha \in J} f_\alpha$, $f_\alpha \in M_\alpha$, for which $\sum_{\alpha \in J} \|f_\alpha\|^2$ is finite: $f_\alpha = 0$ for all $\alpha \in J$ aside from a finite or countable subset and the sum of the non-vanishing elements f_α is convergent.

Proof: The last part of the theorem will be proved first. If $\sum_{\alpha \in J} \|f_\alpha\|^2$ is finite, then the relation $\|f_\alpha\|^2 > \varepsilon$ holds for only a finite set of indices α. If ε is taken successively to be $1, \frac{1}{2}, \frac{1}{3}, \ldots$, then it is evident that $\|f_\alpha\|^2 \neq 0$ and $f_\alpha \neq 0$ for only a finite or countable set of indices α. Let the indices of the non-vanishing elements f_α be denoted by $\alpha_1, \alpha_2, \ldots$. If this sequence is finite, the question of convergence does not arise. If it is infinite, let $f_{\alpha_i} = a_i \varphi_i$, where $a_i = \|f_{\alpha_i}\|$ and $\varphi_i = \frac{1}{\|f_{\alpha_i}\|} f_{\alpha_i}$. The elements $\varphi_1, \varphi_2, \ldots$ form an o.n. set and Theorem 12.16 leads to the result stated.

As $f - f_\alpha = \sum_i f_{\alpha_i} - f_\alpha = \sum_{\alpha_i \neq \alpha} f_\alpha$ (this is obvious if $\alpha = \alpha_i$ for some $i = 1, 2, \ldots$, but it is true for $\alpha \neq \alpha_1, \alpha_2, \ldots$ too, since then $f_\alpha = 0$; in what follows the sequence $\alpha_1, \alpha_2, \ldots$ may be finite or infinite), and as $f_{\alpha_i} \in M_{\alpha_i} \subset \ominus M_\alpha$ for $\alpha_i \neq \alpha$, it follows that $f - f_\alpha \in \ominus M_\alpha$. But $f_\alpha \in M_\alpha$. Hence $P_{M_\alpha} f = f_\alpha$. It is evident that $\|\sum_i a_i \varphi_i\|^2 = \sum_i |a_i|^2$, so that $\|f\|^2 = \|\sum_i f_{\alpha_i}\|^2 = \sum_i \|f_{\alpha_i}\|^2 = \sum_{\alpha \in J} \|f_\alpha\|^2$. This proves part of the first assertion of the theorem: the uniqueness of the elements f_α in the representation $f = \sum_{\alpha \in J} f_\alpha$ for given f, and the explicit formulae concerning them. It remains to prove only the existence of such a representation for an arbitrary f.

If $f \in S$ and if $P_{M_\alpha} f \neq 0$, let $P_{M_\alpha} f = a_\alpha \varphi_\alpha$, where $a_\alpha = \|P_{M_\alpha} f\|$ and $\varphi_\alpha = \frac{1}{\|P_{M_\alpha} f\|} P_{M_\alpha} f$. The elements φ_α form an o.n. set. By Theorem 12,11, Corollary 2, $\|f\|^2 \geq \sum_\alpha |(f, \varphi_\alpha)|^2 = \sum_\alpha |a_\alpha|^2 = \sum_\alpha \|P_{M_\alpha} f\|^2$. The last summation may be extended over all $\alpha \in J$ since $P_{M_\alpha} f = 0$ for those indices α for which φ_α was not defined. Hence $\sum_{\alpha \in J} \|P_{M_\alpha} f\|^2$ is finite and the argument above shows that there exists an element $f' = \sum_{\alpha \in J} P_{M_\alpha} f$ (let $f_\alpha = P_{M_\alpha} f \in M_\alpha$) and $P_{M_\alpha} f' = P_{M_\alpha} f$. Therefore $P_{M_\alpha}(f - f') = 0$, $f - f'$ is orthogonal to M_α for each $\alpha \in J$, $f - f'$ is orthogonal to $[\ldots, M_\alpha, \ldots] = S$ and $f - f' = 0$. This completes the proof.

<u>THEOREM 14.12.</u> <u>Let the sets M_α, $\alpha \in J$, be as in Theorem 14.11. Let A be an operator in S which is linear, closed, s.v., and reduced by each set M_α. By condition 3 of Theorem 14.6, $Af \in M_\alpha$ for $f \in M_\alpha \cdot D(A)$, so that A may be considered as an operator in M_α as long as only elements of M_α are considered; when A is so regarded it will be denoted by A_α; A_α is thus the contraction of A over $M_\alpha \cdot D(A)$. The following assertions are valid:</u>

1) <u>If $f = \sum_{\alpha \in J} f_\alpha$, $f_\alpha \in M_\alpha$ (see Theorem 14.11), then Af is defined if and only all elements $A_\alpha f_\alpha$ are defined and $\sum_{\alpha \in J} \|A_\alpha f\|^2$ is finite; in this event</u>

$$Af = \sum_{\alpha \in J} A_\alpha f_\alpha .$$

2) <u>Each operator A_α is linear, closed, and s.v.</u>

3) <u>Conversely, if in each set M_α there is given a linear, closed, s.v. operator $\bar{A}_\alpha$, there exists a unique linear, closed, s.v. operator A in S which is reduced by each set M_α and for which $A_\alpha = \bar{A}_\alpha$ for all $\alpha \in J$.</u>

Proof: Part 1. Since $\|\langle f, f'\rangle\|^2 = \|f\|^2 + \|f'\|^2$ (see Definition 13.1), part 1 may be reformulated in the following manner: the elements $\langle f, Af\rangle$ coincide with the elements $\sum_{\alpha \in J} \langle f_\alpha, A_\alpha f_\alpha\rangle$, $f_\alpha \in M_\alpha$, where $\sum_{\alpha \in J} \|\langle f_\alpha, A_\alpha f_\alpha\rangle\|^2$ is finite. If Theorem 14.11 is applied to the space G(A) instead of S (see Definition 13.2; G(A) is a c.l.m., and hence is a space to which Theorem 14.11 may be applied), it follows that $G(A) = [\ldots, G(A_\alpha), \ldots]$. But A_α is the contraction of A over $M_\alpha \cdot D(A)$. Hence (in an obvious notation) $G(A_\alpha) = (M_\alpha \times S)\cdot G(A)$. Therefore

$$[\ldots, G(A_\alpha), \ldots] = [\ldots, (M_\alpha \times S)\cdot G(A), \ldots] = ([\ldots, M_\alpha, \ldots] \times S)\cdot G(A) =$$
$$= (S \times S)\cdot G(A) = G(A).$$

Part 2. Since G(A) and M_α are c.l.m.'s, so also are $M_\alpha \times S$ and $G(A_\alpha) = (M_\alpha \times S)\cdot G(A)$. Hence A_α is linear and closed. As A is s.v., its contraction A_α is also s.v.

Part 3. Corresponding to the operators $\bar{A}_\alpha$ in M_α (α ranging over J) there exists a unique operator A satisfying part 1 with $A_\alpha = \bar{A}_\alpha$ inasmuch as it was shown in the proof of part 1 that this condition means merely that $G(A) = [\ldots, G(\bar{A}_\alpha), \ldots]$, and this in turn exactly determines G(A), that is, A . Let this operator A be denoted by $\bar{A}$. Any operator A of the sort described in part 3 would satisfy part 1 with $A_\alpha = \bar{A}_\alpha$; therefore it remains merely to prove that $\bar{A}$ satisfies part 3.

As $G(\bar{A})$ is a c.l.m. by definition, $\bar{A}$ is linear and closed; part 1 shows that it is s.v. (Theorem 14.11). If $f \in D(\bar{A})$, then let $f = \sum_{\alpha \in J} f_\alpha$, $f_\alpha \in M_\alpha$. By application of part 1 to any element f_β ($\beta \in J$) it follows that $f_\beta \in D(\bar{A})$ and $\bar{A}f_\beta = \bar{A}_\beta f_\beta$. But $P_{M_\beta} f = f_\beta$, and as $\bar{A}f = \sum_{\alpha \in J} \bar{A}_\alpha f_\alpha$, $\bar{A}_\alpha f_\alpha \in M_\alpha$. The relation $P_{M_\beta} \bar{A}f = \bar{A}_\beta f_\beta$ is obtained in a similar manner (by Theorem 14.11). Hence $P_{M_\beta} f \in D(\bar{A})$, $\bar{A}P_{M_\beta} f = \bar{A}_\beta f_\beta = P_{M_\beta}\bar{A}f$, so that $\bar{A}$ commutes with P_{M_β} and M_β reduces $\bar{A}$. This condition holds for all $\beta \in J$. It has already been pointed out that the contraction of $\bar{A}$ over $M_\beta \cdot D(\bar{A})$ is $\bar{A}_\beta$. Therefore $\bar{A}$ satisfies part 3, and the proof is complete.

The above decomposition of an operator A in S into operators A_α in M_α is simply related to the various operations which may be performed with A on the basis of the definitions of this and the preceding chapter. These relationships are enumerated in

THEOREM 14.13. *Let the sets M_α, $\alpha \in J$, be as in Theorems 14.11 and 14.12; let A and B be two linear, closed, s.v. operators each of which is reduced by each set M_α; let A_α and B_α be their respective contractions over $M_\alpha \cdot D(A)$ and $M_\alpha \cdot D(B)$* (see Theorem 14.12). *The following assertions are valid:*

1) $A \subset B$ *if and only if* $A_\alpha \subset B_\alpha$ *for every* $\alpha \in J$.
2) A *is partially adjoint to* B *if and only if* A_α *is partially adjoint to* B_α *for every* $\alpha \in J$.
3) $(A^*)_\alpha$ *exists and is equal to* $(A_\alpha)^*$ *for every* $\alpha \in J$.
4) A *has a s.v. inverse if and only if each* A_α *has a s.v. inverse;* $(A^{-1})_\alpha$ *exists and is equal to* $(A_\alpha)^{-1}$ *for every* $\alpha \in J$.
5) *If* K *is any one of the six classes* $K_1, \ldots, K_6$ *enumerated below, then* A *belongs to* K *if and only if* A_α *belongs to* K *for every* $\alpha \in J$.

$K_1, \ldots, K_6$ consist of all operators which are respectively Hermitian, semi-definite, definite, projections, s.a., unitary.

6) *A is bounded if and only if each A_α is bounded and the numbers $|||A_\alpha|||$, $\alpha \in J$, are bounded, in this event,* $|||A||| = \underset{\alpha \in J}{\text{l.u.b.}}\ |||A_\alpha|||$.

Proof: Part 1. The necessity is evident; the sufficiency follows from Theorem 14.12, part 1.

Part 2. Same as part 1 together with the fact that the relation $\|f\|^2 = \sum_{\alpha \in J} \|f_\alpha\|^2$ of Theorem 14.11 may be generalized so as to assume the form $(f, g) = \sum_{\alpha \in J} (f_\alpha, g_\alpha)$ $(f = \sum_{\alpha \in J} f_\alpha$, $g = \sum_{\alpha \in J} g_\alpha$, $f_\alpha \in M_\alpha$, $g_\alpha \in M_\alpha)$. (The real part of this latter relation results from the former by replacing f by $\frac{f+g}{2}$ and $\frac{f-g}{2}$ in succession and taking the difference between the two results; the imaginary part results by replacing f and g respectively by f and ig.)

Part 3. Since A^* is reduced by the sets M_α (Theorem 14.4, part 8), it is possible to form the operators $(A^*)_\alpha$. As A and A^* are partial adjoints, the same is true (by part 2) of A_α and $(A^*)_\alpha$. Hence $(A^*)_\alpha \subset (A_\alpha)^*$. Let A' be the operator such that $A'_\alpha = (A_\alpha)^*$. Then A_α and A'_α are partial adjoints, so that the same is true of A and A'. Hence $A' \subset A^*$, and $(A_\alpha)^* = A'_\alpha \subset A^*_\alpha$ (by part 1). Therefore $(A^*)_\alpha = (A_\alpha)^*$.

Part 4. It must first be shown that the condition $Af = 0$ is equivalent to the condition $A_\alpha f_\alpha = 0$ for every $\alpha \in J$. But the necessity is evident and the sufficiency follows from part 1. Since A^{-1} is reduced by the sets M_α (Theorem 14.4, part 4), it is possible to form the operator $(A^{-1})_\alpha$. It is evident that $(A^{-1})_\alpha = (A_\alpha)^{-1}$.

Part 5. The assertion with regard to K_1 (Hermitian) follows from part 2; with regard to K_5 (s.a.) from part 3; with regard to K_6 (unitary) from parts 3 and 4 together with the fact that, for a unitary operator, $A^{-1} = A^*$

(Remark 3 following Theorem 14.1); with regard to K_4 (projections), the assertion could easily be verified directly, but as a matter of fact it is contained in the assertions regarding K_1 and K_6 since A is a projection if and only if 2A - 1 is Hermitian and unitary (see the discussion at the end of the section dealing with unitary operators); with regard to K_2 (semi-definite), the necessity is evident and the sufficiency follows from the relation $(f, g) = \sum_{\alpha \in J} (f_\alpha, g_\alpha)$ (derived above) since this relation implies the relation $(Af, f) = \sum_{\alpha \in J} (A_\alpha f_\alpha, f_\alpha)$; with regard to K_3 (definite), the assertion follows from this same relation or from K_2 and part 4.

Part 6. It is evident that, if A is bounded, then each A_α is also bounded; in fact, $|||A_\alpha||| \leqq |||A|||$ since each A_α is a contraction of A. Thus the condition is necessary, and $|||A||| \geqq \underset{\alpha \in J}{\text{l.u.b.}}\ |||A_\alpha|||$. Conversely, suppose that each A_α is bounded and that the numbers $|||A_\alpha|||$, $\alpha \in J$, are bounded. Let $C = \underset{\alpha \in J}{\text{l.u.b.}}\ |||A_\alpha|||$. Then $\|A_\alpha g\| \leqq C \cdot \|g\|$ for $g \in M_\alpha$. If in the relation $\|f\|^2 = \sum_{\alpha \in J} \|f\|^2$ the element f is replaced by Af and f_α is replaced by $A_\alpha f_\alpha$, the result is that $\|Af\|^2 = \sum_{\alpha \in J} \|A_\alpha f\|^2$. Hence $\|Af\|^2 \leqq C^2 \cdot \|f\|^2$ and $\|Af\| \leqq C \cdot \|f\|$, that is, A is bounded and $|||A||| \leqq C$. Thus the condition is sufficient and $|||A||| \leqq \underset{\alpha \in J}{\text{l.u.b.}}\ |||A_\alpha|||$. This completes the proof.

Let B be an operator in S; $\mathcal{B}$ is taken to be that operator in $S \times S$ defined by the condition $\mathcal{B}\langle f, g\rangle = \langle Bf, Bg\rangle$. (It is apparent that $\langle f, g\rangle \in D(\mathcal{B})$ if and only if $f \in D(B)$ and $g \in D(B)$.) In the notation of Appendix II, $\mathcal{B}$ is represented by the matrix $\begin{Vmatrix} B & 0 \\ 0 & B \end{Vmatrix}$. If B0 = 0, a straightforward computation shows that $\mathcal{B}^*$ is similarly represented by the matrix $\begin{Vmatrix} B^* & 0 \\ 0 & B^* \end{Vmatrix}$.

Assume that B is s.v. with D(B) = S. The fact that a s.v. operator A commutes with B means that if $f \in D(A)$ then $Bf \in D(A)$ and that ABf = BAf, that is, $\langle Bf, BAf\rangle \in G(A)$, or again, if $\langle f, g\rangle \in G(A)$ then $\langle Bf, Bg\rangle \in G(A)$, or again, if $\varphi \in G(A)$ then $\mathcal{B}\varphi \in G(A)$.

Let $\mathcal{E}_A = P_{G(A)}$. The preceding condition states that $\mathcal{E}_A \mathcal{B}\varphi = \mathcal{B}\varphi$ for all $\varphi \in G(A)$, that is, for all $\varphi = \mathcal{E}_A \psi$. This means that $\mathcal{E}_A \mathcal{B} \mathcal{E}_A \psi = \mathcal{B}\mathcal{E}_A \psi$ for all ψ, or again, $\mathcal{E}_A \mathcal{B} \mathcal{E}_A = \mathcal{B}\mathcal{E}_A$. This leads to

THEOREM 14.14. *If A and B are s.v. with D(B) = S, then A commutes with B when and only when $\mathcal{E}_A \mathcal{B} \mathcal{E}_A = \mathcal{B}\mathcal{E}_A$; if, furthermore, $D(B) = D(B^*) = S$, then A c.a. with B when and only when $\mathcal{E}_A \mathcal{B} = \mathcal{B}\mathcal{E}_A$.*

Proof: The first assertion has just been proved. It implies that in the second assertion the relations $\mathcal{E}_A \mathcal{B} \mathcal{E}_A = \mathcal{B}\mathcal{E}_A$ and $\mathcal{E}_A \mathcal{B}^* \mathcal{E}_A = \mathcal{B}^* \mathcal{E}_A$ constitute a necessary and sufficient condition. As $\mathcal{E}_A$ and $\mathcal{B}$ are bounded and everywhere defined ($\mathcal{B}$ has these properties because B has them) it is possible to apply the operation * to the second relation and obtain the result $\mathcal{E}_A \mathcal{B} \mathcal{E}_A = \mathcal{E}_A \mathcal{B}$. The first and last relations together imply that $\mathcal{E}_A \mathcal{B} = \mathcal{B}\mathcal{E}_A$. Conversely, this relation implies the two just mentioned:

$$\mathcal{E}_A \mathcal{B} \mathcal{E}_A = \mathcal{E}_A \cdot \mathcal{E}_A \mathcal{B} = \mathcal{E}_A \mathcal{B},$$

$$\mathcal{E}_A \mathcal{B} \mathcal{E}_A = \mathcal{B}\mathcal{E}_A \cdot \mathcal{E}_A = \mathcal{B}\mathcal{E}_A.$$

Remark. It should be noted that, in the second part of the preceding theorem, $\mathcal{E}_A$ and $\mathcal{B}$ are both bounded and everywhere defined. Hence the condition $\mathcal{E}_A \mathcal{B} = \mathcal{B}\mathcal{E}_A$ means that $\mathcal{B}$ commutes with $\mathcal{E}_A$ and therefore that $\mathcal{B}$ c.a. with $\mathcal{E}_A$.

By using the matrix notation of Appendix II it is possible to replace $\mathcal{E}_A$ by its matrix $\begin{Vmatrix} A_{11} & A_{12} \\ A_{21} & A_{22} \end{Vmatrix}$. It is evident that $\mathcal{B}$ commutes (c.a.) with $\mathcal{E}_A$ when and only when B commutes (c.a.) with A_{11}, A_{12}, A_{21}, A_{22}. This provides a criterion in S that A and B c.a.

Appendix III.

A familiar way to "arithmetize" operators is to replace them by matrices. However, to do this in the general case (as when the operators $\emptyset$ are linear, closed, s.v., with domains dense in a Hilbert space S) it is necessary to use a certain amount of caution. The purpose of this appendix is to carry out this replacement.

The following theorem is useful to this end:

THEOREM 14'.1. *If M is a separable l.m., then there exists in M a finite or countable o.n. set A:* $\varphi_1, \varphi_2, \ldots$ *such that* $\{A\} \subset M \subset [A]$.

Remark. M need not be closed. If the operation [...] is applied to the relation $\{A\} \subset M \subset [A]$, then [A] = [M] = closure of M. In the case where M = S (S being separable), this result leads to Theorem 12.18.

Proof: Let $f_1, f_2, \ldots$ be a sequence of elements of M which is dense in M. This sequence is also dense in [M] = closure of M. The method of proof of Theorem 12.18 carries over to the space [M] (since M is a c.l.m.) when applied to the sequence $f_1, f_2, \ldots$. The set $\varphi_1, \varphi_2, \ldots$ so obtained satisfies the requirements of the theorem.

It is desirable to consider successively the three essentially distinct possibilities with regard to the dimension Ω of a space S (compare Definition 12.17). Ω may be finite (Ω = N = 0, 1, 2, ...), countably infinite ($\Omega = \omega$), or uncountably infinite ($\Omega > \omega$); correspondingly S will be an N-dimensional Euclidean space, a Hilbert space, or a hyper-Hilbert space.

Case 1. Ω = N = 0, 1, 2, ...; S an N-dimensional Euclidean space.

Let $\emptyset$ be a linear, s.v. operator with D($\emptyset$) dense in S. (In the other two cases it will be necessary to assume closure also, but at present this is unnecessary.) By Theorem 14'.1 there exists an o.n. set A: $\varphi_1, \ldots, \varphi_L$ in

$D(\phi)$ which is dense in $D(\phi)$ and hence also dense in S; it follows that A is complete in S, L = N, and every element of S is of the form $\sum_{\rho=1}^{N} \alpha_\rho \varphi_\rho$, that is, $S = \{A\}$. Since ϕ is linear, $D(\phi) = S$. Thus ϕ is defined over the whole of S, and A may be taken as any complete o.n. set.

Let $\phi \varphi_\rho = \sum_{\sigma=1}^{N} a_{\rho\sigma} \varphi_\sigma$, so that $a_{\rho\sigma} = (\phi \varphi_\rho, \varphi_\sigma)$. If $f = \sum_{\rho=1}^{N} \alpha_\rho \varphi_\rho$ and $\phi f = \sum_{\sigma=1}^{N} y_\sigma \varphi_\sigma$, then $y_\sigma = \sum_{\rho=1}^{N} a_{\rho\sigma} \alpha_\rho$. Furthermore,

$$\| f \|^2 = \sum_{\rho=1}^{N} |\alpha_\rho|^2 , \qquad \| \phi f \|^2 = \sum_{\sigma=1}^{N} \Big| \sum_{\rho=1}^{N} a_{\rho\sigma} \alpha_\rho \Big|^2 ,$$

$$(\phi f, f) = \sum_{\rho,\sigma=1}^{N} a_{\rho\sigma} \alpha_\rho \bar{\alpha}_\sigma .$$

It is evident that the range of values of $\frac{\|\phi f\|}{\|f\|} = \sqrt{\frac{\sum_{\sigma=1}^{N} \left| \sum_{\rho=1}^{N} a_{\rho\sigma} \alpha_\rho \right|^2}{\sum_{\rho=1}^{N} |\alpha_\rho|^2}}$ is bounded since N is finite and constant. By Definition 13.16 the operator ϕ is bounded; since $D(\phi) = S$, ϕ is closed. By Theorem 13.21 or 13.23, ϕ^* exists and has the same character as ϕ; if $a'_{\rho\sigma} = (\phi^* \varphi_\rho, \varphi_\sigma)$, then $a'_{\rho\sigma} = (\varphi_\rho, \phi \varphi_\sigma) = \overline{a_{\sigma\rho}}$. These results are summarized in

<u>THEOREM 14'.2.</u> <u>If ϕ is linear, s.v., with $D(\phi)$ dense in S, then ϕ is everywhere defined, bounded, and closed; ϕ^* exists and has the same properties.</u>

<u>If A is a complete o.n. set $\varphi_1, \ldots, \varphi_N$, form the representations $\phi \varphi_\rho = \sum_{\sigma=1}^{N} a_{\rho\sigma} \varphi_\sigma$ so that $a_{\rho\sigma} = (\phi \varphi_\rho, \varphi_\sigma)$. Then the matrix $\| a_{\rho\sigma} \|$ is said to belong to the operator ϕ for the set $A = (\varphi_1, \ldots, \varphi_N)$.</u>

<u>In the same sense, the matrix $\| a'_{\rho\sigma} \|$, where $a'_{\rho\sigma} = \overline{a_{\sigma\rho}}$, will belong to ϕ^* for the set $A = (\varphi_1, \ldots, \varphi_N)$.</u>

<u>If $f = \sum_{\rho=1}^{N} x_\rho \varphi_\rho$ and $\phi f = \sum_{\sigma=1}^{N} y_\sigma \varphi_\sigma$ then $y_\sigma = \sum_{\rho=1}^{N} a_{\rho\sigma} \alpha_\rho$.</u>

This theorem (proved above) leads directly to the following

Corollary: *Since* $D(\phi) = S$ *and since* ϕ^* *is s.v. (Theorem 13.14), the condition* $\phi \subset \phi^*$ *implies the condition* $\phi = \phi^*$, *that is, if* ϕ *is Hermitian then it is s.a. (Theorem 13.18). Furthermore, the condition* $\phi = \phi^*$ *is equivalent to* $a_{\rho\sigma} = a'_{\rho\sigma}$ *for all* ρ, σ *(since* $a_{\rho\sigma}$ $(a'_{\rho\sigma})$ *determines* $\phi(\phi^*)$*), that is* $a_{\rho\sigma} = \overline{a_{\sigma\rho}}$.

Case 2. $\Omega = \omega$; S a Hilbert space.

Let ϕ be a linear, closed, s.v. operator with $D(\phi)$ dense in S. By Theorem 14'.1 there exists an o.n. set A: $\varphi_1, \varphi_2, \ldots$ in $D(\phi)$ which is complete in S. Let ϕ_1 be the contraction of ϕ over A, that is, $D(\phi_1) = A$, $\phi_1 \subset \phi$. Since A is complete in S it is possible to introduce the representation $\phi\varphi_\rho = \phi_1 \varphi_\rho = \sum_{\sigma=1}^{\infty} a_{\rho\sigma}\varphi_\sigma$, where $a_{\rho\sigma} = (\phi\varphi_\rho, \varphi_\sigma) = (\phi_1\varphi_\rho, \varphi_\sigma)$. Since $\phi_1 \subset \phi$ and since ϕ is linear and closed, it follows that $\tilde{\phi}_1 \subset \phi$.

THEOREM 14'.3. *If* ϕ *is linear, closed, s.v., with* $D(\phi)$ *dense in* S, *then there exists an o.n. set* A: $\varphi_1, \varphi_2, \ldots$ *in* $D(\phi)$ *which is complete in* S *and such that* $\tilde{\phi}_1 = \phi$, *where* ϕ_1 *is the contraction of* ϕ *over* A.

Remark: If $D(\phi) = S$ and if ϕ is bounded (so that ϕ is continuous over S), then ϕ_1 is continuous over $D(\phi_1)$ since $\tilde{\phi}_1 \subset \phi$. Hence $D(\phi_1)$ is a c.l.m. (Theorem 13.11). But $D(\tilde{\phi}_1) \supset D(\phi_1) = A$ so that $D(\tilde{\phi}_1) \supset [A] = S$. Hence $D(\tilde{\phi}_1) = S$ and $\tilde{\phi}_1 = \phi$. Under these conditions every complete o.n. set A satisfies the conditions of the theorem. It will be shown in a later chapter that in all other cases not every complete o.n. set in $D(\phi)$ satisfies the conditions of the theorem.

Proof: Since S is separable, $S \times S$ and $G(\phi) \subset S \times S$ are also separable. Hence there exists a sequence $\langle f_1, \phi f_1\rangle, \langle f_2, \phi f_2\rangle, \ldots$ dense in $G(\phi)$. If $\langle f_{n_1}, \phi f_{n_1}\rangle, \langle f_{n_2}, \phi f_{n_2}\rangle, \ldots$ is a subsequence of this sequence with limit

$\langle f, f^*\rangle$, then, since $\emptyset$ is closed, $\langle f, f^*\rangle \varepsilon G(\emptyset)$, that is, $f \varepsilon D(\emptyset)$ and $\emptyset f = f^*$. If $\emptyset_o$ is the contraction of $\emptyset$ over the set B: $f_1, f_2, \ldots$, then $\tilde{\emptyset}_o = \emptyset$.

It is evident that the set B is dense in $D(\emptyset)$ and S. Let A be the o.n. set $\varphi_1, \varphi_2, \ldots$ arising from $M = \{f_1, f_2, \ldots\} = \{B\}$ according to the constructions of Theorems 12.18 and 14'.1. Since $\{A\} = \{B\}$, it follows that $\hat{\emptyset}_1 = \hat{\emptyset}_o$, $\tilde{\emptyset}_1 = \tilde{\emptyset}_o = \emptyset$; and $[A] = [B] = S$. Hence A satisfies the conditions of the theorem.

Definition 14'.1. A complete o.n. set A: $\varphi_1, \varphi_2, \ldots$ which satisfies the conditions of the preceding theorem will be called a determining set for $\emptyset$.

Since A is complete, the relations $\emptyset\varphi_\rho = \emptyset_1\varphi_\rho = \sum_{\sigma=1}^{\infty} a_{\rho\sigma}\varphi_\sigma$ and $a_{\rho\sigma} = (\emptyset\varphi_\rho, \varphi_\sigma) = (\emptyset_1\varphi_\rho, \varphi_\sigma)$ noted above still obtain. The matrix $\|a_{\rho\sigma}\|$ is said to belong to the operator $\emptyset$ for the $A = (\varphi_1, \varphi_2, \ldots)$. (These a's are closely analogous to the a's used in Case 1.)

By Theorem 12.16, $\sum_{\sigma=1}^{\infty}|a_{\rho\sigma}|^2 = \sum_{\sigma=1}^{\infty}|(\emptyset\varphi_\rho, \varphi_\sigma)|^2 = \|\emptyset\varphi_\rho\|^2$ is finite.

It is evident that $D(\hat{\emptyset}_1) = \{D(\emptyset_1)\} = \{A\}$, that is, $D(\hat{\emptyset}_1)$ is the set of all elements $f = \sum_{\rho=1}^{n} x_\rho\varphi_\rho$, where n is any positive integer. If $\emptyset f = \tilde{\emptyset}_1 f = \sum_{\sigma=1}^{\infty} y_\sigma\varphi_\sigma$ then $\hat{\emptyset}_1 f = \sum_{\rho=1}^{n} x_\rho\emptyset_1\varphi_\rho = \sum_{\rho=1}^{n} x_\rho \sum_{\sigma=1}^{\infty} a_{\rho\sigma}\varphi_\sigma = \sum_{\sigma=1}^{\infty}\left(\sum_{\rho=1}^{n} a_{\rho\sigma}x_\rho\right)\varphi_\sigma$, and $y_\sigma = \sum_{\rho=1}^{n} a_{\rho\sigma}x_\rho$.

Although it is not possible to give a direct characterization of $D(\tilde{\emptyset}_1) = D(\emptyset)$ (except by directly translating Definition 13.10 and the discussion following it), something can be said in the event that $A \subset D(\emptyset^*)$, i.e. $\emptyset^*\varphi_\sigma$ is defined for all φ_σ in A. By Theorem 12.8, $(\emptyset f, \varphi_\sigma) = (f, \emptyset^*\varphi_\sigma)$ is a continuous function of f over $D(\emptyset)$. Hence if $f \varepsilon D(\emptyset)$, $f = \sum_{\rho=1}^{\infty} x_\rho\varphi_\rho$, $\emptyset f = \sum_{\sigma=1}^{\infty} y_\sigma\varphi_\sigma$, then $y_\sigma = (\emptyset f, \varphi_\sigma)$ is a continuous function of f. Let

$f^{(n)} = \sum_{\rho=1}^{n} x_\rho \varphi_\rho$, $\phi f^{(n)} = \sum_{\sigma=1}^{\infty} y_\sigma^{(n)} \varphi_\sigma$; then $f = \lim_{n\to\infty} f^{(n)}$, $y_\sigma = \lim_{n\to\infty} y_\sigma^{(n)} =$
$= \lim_{n\to\infty} \sum_{\rho=1}^{n} a_{\rho\sigma} x_\rho = \sum_{\rho=1}^{\infty} a_{\rho\sigma} x_\rho$.

In this discussion it was proved incidentally that $\sum_{\rho=1}^{\infty} a_{\rho\sigma} x_\rho$ is convergent. However, if $f = \sum_{\rho=1}^{\infty} x_\rho \varphi_\rho$ is an arbitrary element of S, this series is even absolutely convergent, the reason being that

$$\sum_{\rho=1}^{\infty} |a_{\rho\sigma}|^2 = \sum_{\rho=1}^{\infty} |(\phi \varphi_\rho, \varphi_\sigma)|^2 = \sum_{\rho=1}^{\infty} |(\varphi_\rho, \phi^* \varphi_\sigma)|^2 = \|\phi^* \varphi_\sigma\|^2$$

and $\sum_{\rho=1}^{\infty} \|x_\rho\|^2 = \|f\|^2$, so that $\sum_{\rho=1}^{\infty} |a_{\rho\sigma}|^2$ and $\sum_{\rho=1}^{\infty} |x_\rho|^2$ are both finite, while $|a_{\rho\sigma} x_\rho| \leq \frac{1}{2}|a_{\rho\sigma}|^2 + \frac{1}{2}|x_\rho|^2$.

<u>Definition 14'.2. If ϕ, A: $\varphi_1, \varphi_2, \ldots$, and $a_{\rho\sigma} = (\phi \varphi_\rho, \varphi_\sigma)$ are as above with $A \subset D(\phi^*)$, then let $f = \sum_{\rho=1}^{\infty} x_\rho \varphi_\rho$ range through S and form the expressions $y_\sigma = \sum_{\sigma=1}^{\infty} a_{\rho\sigma} x_\rho$. Let ϕ_2 be that operator whose domain $D(\phi_2)$ is the set of all elements f such that $\sum_{\sigma=1}^{\infty} |y_\sigma|^2$ is finite and which is defined by the condition $\phi_2 f = \sum_{\sigma=1}^{\infty} y_\sigma \varphi_\sigma$, $f \in D(\phi_2)$.</u>

It follows from the preceding discussion that $\phi \subset \phi_2$.

It is possible to characterize ϕ_2 completely in terms of ϕ. Since $A \subset D(\phi^*)$, it is possible to define $\phi^{*'}$ as the contraction of ϕ^* over A. The conditions $f \in D(\phi^{*'})^*$ and $f^* = (\phi^{*'})^* f$ together mean that $(f, \phi^{*'} g) = (f^*, g)$ for all elements g in $D(\phi^{*'}) = A$, i.e., for all elements $g = \varphi_\sigma$. Let $f = \sum_{\rho=1}^{\infty} x_\rho \varphi_\rho$, $f^* = \sum_{\sigma=1}^{\infty} z_\sigma \varphi_\sigma$ so that $(f^*, \varphi_\sigma) = z_\sigma$ and $(f, \phi^{*'} \varphi_\sigma) =$
$= (f, \phi^* \varphi_\sigma) = \sum_{\rho=1}^{\infty} (f, \varphi_\rho)(\varphi_\rho, \phi^* \varphi_\sigma) = \sum_{\rho=1}^{\infty} (f, \varphi_\rho)(\phi \varphi_\rho, \varphi_\sigma) = \sum_{\rho=1}^{\infty} a_{\rho\sigma} x_\rho =$
$= y_\sigma$. Hence $y_\sigma = z_\sigma$. Since the numbers z_σ are arbitrary except for the

condition that $\sum_{\sigma=1}^{\infty} |z_\sigma|^2$ be finite, it follows that $f^* = \sum_{\sigma=1}^{\infty} y_\sigma \varphi_\sigma$, where $\sum_{\sigma=1}^{\infty} |y_\sigma|^2$ is finite, i.e., that $(\emptyset^{*'}) = \emptyset_2$.

Inasmuch as $\emptyset^{*'} \subset \emptyset^*$, $(\emptyset^{*'})^* \supset \emptyset^{**}$ and $\emptyset_2 \supset \emptyset$ (Theorem 13.13 and the remark preceding it). But $\emptyset_2 = (\emptyset^{*'})^* = (\widetilde{\emptyset^{*'}})^*$ (cf. l.c.), so that the condition $\emptyset_2 = \emptyset$ is equivalent to the condition $(\widetilde{\emptyset^{*'}})^* = \emptyset^{**}$ ($\emptyset$ is linear and closed so that $\emptyset = \emptyset^{**}$ (Theorem 13.13)); as $\widetilde{\emptyset}^{*'}$ and $\emptyset^*$ are both linear and closed, each of these conditions is equivalent to the condition $\widetilde{\emptyset}^{*'} = \emptyset^*$. But this is obviously equivalent to the statement that the complete o.n. set **A** (assumed contained in $D(\emptyset^*)$) is a determining set for $\emptyset^*$. These results are summarized in

THEOREM 14'.4. (i) *Let $\emptyset$ be linear, closed, s.v., with $D(\emptyset)$ dense in S; let A: $\varphi_1, \varphi_2, \ldots$ be a complete o.n. set determining $\emptyset$; let $\emptyset_1$ be the contraction of $\emptyset$ over $A \subset D(\emptyset)$; introduce the representation $\emptyset \varphi_\rho = \emptyset_1 \varphi_\rho = \sum_{\sigma=1}^{\infty} a_{\rho\sigma} \varphi_\sigma$ so that $a_{\rho\sigma} = (\emptyset \varphi_\rho, \varphi_\sigma) = (\emptyset_1 \varphi_\rho, \varphi_\sigma)$. Then*

(1) $\sum_{\sigma=1}^{\infty} |a_{\rho\sigma}|^2 = \|\emptyset \varphi_\rho\|^2$ *is finite.*

(2) *$D(\emptyset_1)$ is the set of all elements $f = \sum_{\rho=1}^{n} x_\rho \varphi_\rho$, and $\emptyset f = \emptyset_1 f = \sum_{\sigma=1}^{\infty} y_\sigma \varphi_\sigma$ with $y_\sigma = \sum_{\rho=1}^{n} a_{\rho\sigma} x_\rho$.*

(3) *While $\widetilde{\emptyset}_1 = \emptyset$, the only way to characterize it is to apply to it Definition 13.10 and the discussion following this definition.*

(ii) *However, if $A \subset D(\emptyset^*)$ ($\emptyset^*$ is linear, closed, s.v., with $D(\emptyset^*)$ dense in S; see Theorem 13.23), then*

(1') $\sum_{\rho=1}^{\infty} |a_{\rho\sigma}|^2 = \|\emptyset^* \varphi_\sigma\|^2$ *is finite.*

(2') *Let $\emptyset^{*'}$ be the contraction of $\emptyset^*$ over $A \subset D(\emptyset^*)$, and let $\emptyset_2$ be the operator introduced in Definition 14'.2. Then $\emptyset_2 = (\emptyset^{*'})^*$, implying that $\emptyset \subset \emptyset_2$; these conditions in turn imply that $\emptyset_2$ is linear, closed,* s.v., with $D(\emptyset_2)$ dense in S.

(3') The conditions $\emptyset = \emptyset_2$ and $\widetilde{\emptyset}^{*'} = \emptyset^*$ are equivalent with each other, and with the requirement that A is a determining set for $\emptyset^*$.

This theorem (proved above) leads directly to

Corollary 1. If $D(\emptyset) = S$, then both $\emptyset$ and $\emptyset^*$ are everywhere defined and bounded (Theorems 13.22 and 13.21). Hence every set A is a determining set for both $\emptyset$ and $\emptyset^*$. In this case Part 2' of the preceding theorem characterizes $\emptyset = (\emptyset_2)$, and is even simplified by the fact that $D(\emptyset) = S$.

Corollary 2. Since $\emptyset = \widetilde{\emptyset}_1 = \emptyset_1^{**}$ (Theorems 14'.3 and 13.13), it follows by Theorem 13.16 that $\emptyset$ is Hermitian when and only when $\emptyset_1$ is Hermitian. Hence, for all ρ, σ, $(\emptyset_1\varphi_\rho, \varphi_\sigma) = (\varphi_\rho, \emptyset_1\varphi_\sigma)$, that is, $a_{\rho\sigma} = \overline{a_{\sigma\rho}}$.

Unfortunately, no comparably simple criterion is known for the property of being s.a.

Corollary 3. If $\emptyset$ is Hermitian, then $\emptyset \subset \emptyset^*$, so that $A \subset D(\emptyset) \subset D(\emptyset^*)$. Hence the assumption in (ii) of the preceding theorem is satisfied. Since $\emptyset_1 \subset \emptyset \subset \emptyset^*$, it is evident that $\emptyset^{*'} = \emptyset_1$. Therefore $\emptyset_2 = (\emptyset^{*'})^* = \emptyset_1^* = \widetilde{\emptyset}_1^* = \emptyset^*$. The condition $\emptyset \subset \emptyset_2$ in 2' assumes the well-known form $\emptyset \subset \emptyset^*$; if $\emptyset = \emptyset_2$, then $\emptyset = \emptyset^*$ and $\emptyset$ is s.a. Hence, by 3', A is a determining set for $\emptyset^*$ if and only if $\emptyset$ is s.a.

It is possible to follow the procedure inverse to that outlined above: start with a complete o.n. set A: $\varphi_1, \varphi_2, \ldots$ and a matrix $\|a_{\rho\sigma}\|$, and construct the operators $\emptyset, \emptyset^*, \emptyset_1, \emptyset_2$.

THEOREM 14'.5. 1) Let A be a complete o.n. set $\varphi_1, \varphi_2, \ldots$ and let $\|a_{\rho\sigma}\|$ be a matrix (of complex numbers). There exists a linear, closed, s.v. operator $\emptyset$, with $D(\emptyset)$ dense in S, to which $\|a_{\rho\sigma}\|$ belongs for A (Definition 14'.1) if and only if every series $\sum_{\sigma=1}^{\infty} |a_{\rho\sigma}|^2$ is finite, and in this event $\emptyset$ is unique.

2) The condition $A \subset D(\emptyset^*)$ obtains if and only if every series $\sum_{\rho=1}^{\infty} |a_{\rho\sigma}|^2$ is finite.

Proof of 1): The necessity of the finiteness of every series $\sum_{\sigma=1}^{\infty} |a_{\rho\sigma}|^2$ follows from Theorem 14'.4, 1); the uniqueness of $\emptyset$ (if it exists at all) follows from the uniqueness of $\emptyset_1$ Theorem 14'.4, 2) and the fact that $\emptyset = \widetilde{\emptyset}_1$. As to the sufficiency of the condition that every series $\sum_{\sigma=1}^{\infty} |a_{\rho\sigma}|^2$ be finite: it is possible to define an operator P over A by the condition $P\varphi_\rho = \sum_{\sigma=1}^{\infty} a_{\rho\sigma}\varphi_\sigma$ and to let $\emptyset = \widetilde{P}$. It is evident that $A \subset D(\emptyset)$, $\emptyset_1 = P$, and $\emptyset = \widetilde{P} = \widetilde{\emptyset}_1$, so that A is a determining set for $\emptyset$; furthermore, $\emptyset\varphi_\rho = P\varphi_\rho = \sum_{\rho=1}^{\infty} a_{\rho\sigma}\varphi_\sigma$.

Proof of 2): The necessity of the finiteness of every series $\sum_{\rho=1}^{\infty} |a_{\rho\sigma}|^2$ follows from Theorem 14'.4,1'); to show this condition sufficient, let $\varphi_\sigma^* = \sum_{\rho=1}^{\infty} a_{\rho\sigma}\varphi_\rho$. Then $(\emptyset_1\varphi_\rho, \varphi_\sigma) = a_{\rho\sigma} = (\varphi_\rho, \varphi_\sigma^*)$, so that $\varphi_\sigma \varepsilon D(\emptyset_1^*)$; but $\emptyset_1^* = \widetilde{\emptyset}_1^* = \emptyset^*$.

The theory of operators, as originally formulated by Hilbert and as developed until recent times, was usually based on matric representation in a fixed orthogonal system. This method is quite convenient for operators which are defined everywhere and bounded. (Most of Hilbert's theory was restricted to operators of this type.) But it is evident that for unbounded operators the description becomes rather involved. It will be seen in later chapters that a number of characteristically "pathological" results may be obtained for unbounded operators (which represent the "general case"). The complicated and "pathological" behavior of matrices representing unbounded operators is a strong argument in favor of the present abstract method. Of course this does not lessen the importance of learning as much as possible about the behavior of matrices of operators.

It should be emphasized that the conditions prevailing in Case 1 (finite dimensional Euclidean spaces) as described by Theorem 14'.2 may be considered as highly simplified and special cases of those conditions in Case 2 (Hilbert space), the chief element of this simplification being that in Case 1 every linear manifold is closed and every linear operator is closed and bounded.

Case 3. $\Omega > \omega$; S a hyper-Hilbert space.

Let ϕ be a linear, closed, s.v. operator with $D(\phi)$ dense in S (compare Theorem 13.23)). By application of Theorem 14.10 and the remark following it, there is obtained a set J of indices α and for each $\alpha \in J$ a Hilbert space M_α such that 1) if $\alpha \neq \beta$, then M_α and M_β are orthogonal, 2) $[\ldots, M_\alpha, \ldots] = S$, where α ranges through J, and 3) each M_α reduces ϕ.

Since the dimension of M_α is ω and since $[\ldots, M_\alpha, \ldots] = S$, it follows that (power of J)$\cdot\,\omega = \Omega$. As Ω is non-countably infinite, the power of J must be infinite. Hence (power of J)$\cdot\,\omega$ = (power of J), so that the power of J is Ω. Thus J is non-countably infinite. (In Case 2, where $\Omega = \omega$, it would be possible for the power of J to be unity.)

By Theorems 14.11 and 14.12, the operator ϕ is completely characterized in S by its behavior in each space M_α, that is, by its contractions ϕ_α over $M_\alpha \cdot D(\phi)$ (α ranging through J). But the operators ϕ_α are in the Hilbert spaces M_α, so that the results obtained in Case 2 are applicable to them. By making this application it is found that the behavior of ϕ in Case 3 is quite analogous to that in Case 2. The precise formulation of the situation is as follows:

Let I be a set (of indices) of power Ω which can be used to label the complete o.n. sets of S. (S is homeomorphic to the space H_I; compare Definition 12.16 and Theorems 12.26 and 12.27.) It will be assumed for the moment that I is non-countably infinite (characteristic of Case 3), but it will be seen later that this assumption is unnecessary .

THEOREM 14'.6. *If $\emptyset$ is a linear, closed, s.v. operator with $D(\emptyset)$ dense in S, then there exists an o.n. set A: φ_α, $\alpha \in I$, in $D(\emptyset)$ which is complete in S and such that $\tilde{\emptyset}_1 = \emptyset$, where $\emptyset_1$ is the contraction of $\emptyset$ over A.*

Remark: Exactly the same remark (with exactly the same proof) can be made here as following Theorem 14'.3.

Proof: Let M_α , $\alpha \in J$, be the Hilbert spaces of Theorem 14.10 as described above; let $\emptyset_\alpha$ be the contractions of $\emptyset$ over $M_\alpha \cdot D(\emptyset)$. By application of Theorem 14'.3 to $\emptyset_\alpha$ in M_α , there exists an o.n. set $A_\alpha : \psi_{\alpha,1}, \psi_{\alpha,2}, \ldots$ in $D(\emptyset_\alpha)$ which is complete in M_α and which is a determining set for $\emptyset_\alpha$.

The entire set A: $\psi_{\alpha,n}$, $\alpha \in J$, n = 1, 2, ..., is o.n.; but $[A] \supset [A_\alpha] = M_\alpha$, $[A] \supset [\ldots, M_\alpha , \ldots] = S$, so that A is complete in S. Since $A_\alpha \subset D(\emptyset_\alpha) \subset D(\emptyset)$, it follows that $A \subset D(\emptyset)$. Let $\emptyset_1$ be the contraction of $\emptyset$ over $A \cdot D(\emptyset)$, and let $\emptyset_{\alpha 1}$ be the contraction of $\emptyset_\alpha$ over $A_\alpha \cdot D(\emptyset_\alpha) = A_\alpha \cdot D(\emptyset) = A_\alpha \cdot D(\emptyset_1)$. Then $(\tilde{\emptyset}_1)_\alpha \supset (\emptyset_\alpha)_1$, and hence $(\tilde{\emptyset}_1)_\alpha \supset \widetilde{(\emptyset_\alpha)}_1 = \emptyset$ (since A_α is a determining set for $\emptyset_\alpha$). By Theorem 14.13,1), it follows that $\tilde{\emptyset}_1 \supset \emptyset$; but it is obvious that $\tilde{\emptyset}_1 \subset \emptyset$, so that $\tilde{\emptyset}_1 = \emptyset$. Therefore, to complete the proof of the theorem it remains merely to change the labeling from A: $\psi_{\alpha,n}$, $\alpha \in J$, n = 1, 2, ..., into A: φ_α, $\alpha \in I$.

Definition 14'.3. *A complete o.n. set A: φ_α, $\alpha \in I$, which satisfies the conditions of the preceding theorem will be called a determining set for $\emptyset$.*

Since A is complete, it is again possible to introduce the representations $\emptyset \varphi_\alpha = \emptyset_1 \varphi_\alpha = \sum_\beta a_{\alpha\beta} \varphi_\beta$, $\alpha \in I$, $\beta \in I$, so that $a_{\alpha\beta} = (\emptyset \varphi_\alpha, \varphi_\beta) = (\emptyset_1 \varphi_\alpha, \varphi_\beta)$. The matrix $\| a_{\alpha\beta} \|$ is said to belong to the operator $\emptyset$ for the set $A = (\ldots, \varphi_\alpha , \ldots)$.

THEOREM 14'.7. (i) Let ϕ be linear, closed, s.v., with $D(\phi)$ dense in S; let A: φ_α, $\alpha \in I$, be a complete o.n. set determining ϕ; let ϕ_1 be the contraction of ϕ over $A \subset D(\phi)$; introduce the representation $\phi\varphi_\alpha = \phi_1\varphi_\alpha = \sum_\beta a_{\alpha\beta}\varphi_\beta$, $\alpha \in I$, $\beta \in I$, so that $a_{\alpha\beta} = (\phi\varphi_\alpha, \varphi_\beta) = (\phi_1\varphi_\alpha, \varphi_\beta)$. Then

1) $\sum_{\beta \in I} |a_{\alpha\beta}|^2 = \|\phi\varphi_\alpha\|^2$ is finite (so that, for any particular α, $a_{\alpha\beta} = 0$ for all β's except for at most a countable set which depends on α).

2) $D(\phi_1)$ is the set of all elements $f = \sum_{\alpha \in L} x_\alpha\varphi_\alpha$, where L is any finite subset of I, and $\phi f = \hat{\phi}_1 f = \sum_{\beta \in I} y_\beta\varphi_\beta$ with $y_\beta = \sum_{\alpha \in L} a_{\alpha\beta}x_\alpha$.

3) While $\tilde{\phi}_1 = \phi$, the only way to characterize it is to apply to it Definition 13.10 and the discussion following this definition.

(ii) However, if $A \subset D(\phi^*)$ (ϕ^* is linear, closed, s.v., with $D(\phi^*)$ dense in S; see Theorem 13.23), then

1') $\sum_{\alpha \in I} |a_{\alpha\beta}|^2 = \|\phi^*\varphi_\beta\|^2$ is finite (so that, for any particular β, $a_{\alpha\beta} = 0$ for all α's except for at most a countable set which depends on β).

2'). Let $\phi^{*'}$ be the contraction of ϕ^* over $A \subset D(\phi^*)$, and let ϕ_2 be the operator defined as follows: let $f = \sum_{\alpha \in I} x_\alpha\varphi_\alpha$ range through S and form the expressions $y_\beta = \sum_{\alpha \in I} a_{\alpha\beta}x_\alpha$ (these sums are all absolutely convergent). Then $D(\phi_2)$ is the set of all elements f such that $\sum_{\beta \in I} |y_\beta|^2$ is finite and, for $f \in D(\phi_2)$, $\phi_2 f$ is taken to be $\sum_{\beta \in I} y_\beta\varphi_\beta$. It follows that $\phi_2 = (\phi^{*'})^*$ implying that $\phi \subset \phi_2$; these conditions in turn imply that ϕ_2 is linear, closed, s.v., with $D(\phi_2)$ dense in S.

3') The conditions $\phi = \phi_2$ and $\widetilde{\phi^{*'}} = \phi^*$ are equivalent, so that A is a determining set for ϕ^*.

Remark concerning 2'). If $\sum_{\alpha \in I} |x_\alpha|^2$ is finite, $x_\alpha = 0$ for all except at most a countable set, and for any particular α in this set, $a_{\alpha\beta} = 0$ for all β except for at most a countable set. Hence if $y_\beta = \sum_{\alpha \varepsilon I} a_{\alpha\beta} x_\alpha$, an at most countable set of y's is distinct from 0. Thus the condition that $\sum_{\beta \varepsilon I} |y_\beta|^2$ be finite is not quite as strong as it might seem.

Proof: Exactly the same as for Theorem 14'.4, where the set (1, 2,...) is always replaced by the set I, and where the sets (1, ..., n) are replaced by finite subsets L of I.

Corollaries 1-3. Exactly the same as for Theorem 14'.4, with exactly the same proof.

THEOREM 14'.8. 1) Let A be a complete o.n. set φ_α, $\alpha \varepsilon$ I, and let $\| a_{\alpha\beta} \|$ be a matrix (of complex numbers) with $\alpha \varepsilon$ I, $\beta \varepsilon$ I. There exists a linear, closed, s.v. operator $\emptyset$, with D($\emptyset$) dense in S, to which a belongs for A (Definition 14'.3) if and only if every series $\sum_{\alpha \varepsilon I} |a_{\alpha\beta}|^2$ is finite, and in this event $\emptyset$ is unique.

2) The condition A $\subset$ D($\emptyset^*$) obtains if and only if every series $\sum_{\beta \varepsilon I} |a_{\alpha\beta}|^2$ is finite.

Proof: Exactly the same as for Theorem 14'5, where the set (1,2,...) is always replaced by the set I.

By comparing the results of Cases 2 and 3 it is seen that they are identical when, in Case 3, I is countable and unessentially specialized to I = (1, 2, ...)). In fact, the results of Case 2 follow directly from Case 3: the non-countability of J was not utilized, and Case 2 arises by taking J = (1) with M_1 = S. Case 1 arises for a finite I, and its conditions, described by Theorem 14'.2, also satisfy the theorems in Case 3 of which they are extremely simplified special cases (see the discussion at the end of Case 2).

In brief, the results of Case 3 apply to any S and any I. However, if S is N-dimensional Euclidean (I finite, Case 1), then essentially more is known (cf. Theorem 14.2 and the discussion at the end of Case 2).

Milton Keynes UK
Ingram Content Group UK Ltd.
[illegible]
444351UK00003B/8

9 780691 095790